ARISTOTLE ON TIME

Aristotle's definition of time as "a number of motion with respect to the before and after" has been branded as patently circular by commentators ranging from Simplicius to W. D. Ross. In this book Tony Roark presents an interpretation of the definition that renders it not only non-circular, but also worthy of serious philosophical scrutiny. He shows how Aristotle developed an account of the nature of time that is inspired by Plato while also thoroughly bound up with Aristotle's sophisticated analyses of motion and perception. When Aristotle's view is properly understood, Roark argues, it is immune to devastating objections against the possibility of temporal passage articulated by McTaggart and other twentieth-century philosophers. Roark's novel and fascinating interpretation of Aristotle's temporal theory will appeal to those interested in Aristotle, ancient philosophy, and the philosophy of time.

TONY ROARK is Associate Professor of Philosophy at Boise State University. His work has appeared in journals such as *Mind*, *Ancient Philosophy*, *Apeiron*, and *History and Philosophy of Logic*.

ARISTOTLE ON TIME

A Study of the Physics

TONY ROARK

Boise State University

CAMBRIDGE
UNIVERSITY PRESS

CAMBRIDGE UNIVERSITY PRESS
Cambridge, New York, Melbourne, Madrid, Cape Town, Singapore,
São Paulo, Delhi, Dubai, Tokyo, Mexico City

Cambridge University Press
The Edinburgh Building, Cambridge CB2 8RU, UK

Published in the United States of America by Cambridge University Press, New York

www.cambridge.org
Information on this title: www.cambridge.org/9781107002623

First published 2011

Printed in the United Kingdom at the University Press, Cambridge

A catalogue record for this publication is available from the British Library

Library of Congress Cataloguing in Publication data
Roark, Tony.
Aristotle on Time : A Study of the Physics / Tony Roark.
p. cm
Includes bibliographical references and indexes.
ISBN 978-1-107-00262-3
1. Aristotle. 2. Time. I. Title.
B491.T5R53 2011
115.092–dc22
2010049731

ISBN 978-1-107-00262-3 Hardback

For Keenan,
whose four-year-old syllogisms about crocodiles and human animals were no less endearing for their invalidity.

Contents

Preface

"It's about time." These words were inscribed in a card given to me by my cohorts and professors upon successfully clarifying my dissertation proposal in 1996. The inscription was intended as a bit of good-natured ribbing and word play, but it was prophetic, for this book has taken far too long to write. It began its life as my doctoral dissertation, written at the University of Washington under the capable and supportive supervision of Professors S. Marc Cohen, David Keyt, Jean Roberts, and Cass Weller. I defended the dissertation in 1999, and since that time I have subjected it to prolonged and radical surgery: superfluous sections have been excised, essential but missing arguments and analyses have been added, and the remaining original material has been improved (or so I hope) in various ways.

Even though the final product bears a weak resemblance to its original incarnation, I owe a deep debt of gratitude to the individuals mentioned above, not only for their help in writing the dissertation, but also for their unwavering kindness and friendship.

I am profoundly grateful for the material support of the National Endowment for the Humanities, which provided a Summer Stipend in 2005 to work toward the completion of this book.[1] I wish I could have lived up to my optimistic projections for finishing it around that time, but I am confident that it is better for the extra attention it has received in the interim.

The Division of Research and the College of Arts and Sciences at Boise State University provided additional material support for my research in the form of a Faculty Research Associates Program Grant, also awarded in 2005.

[1] Any views, findings, conclusions, or recommendations expressed in this publication do not necessarily reflect those of the National Endowment for the Humanities.

Several parts of this book have been published previously in different form as journal articles, and I wish to extend my gratitude to the editors of those journals for granting permission to reprint that material here. In particular, portions of Chapter 3 were published under the title "Why Aristotle Says There Is No Time without Change" in *Apeiron*;[2] parts of Chapters 4 and 5 appeared in *Ancient Philosophy* under the title "Aristotle's Definition of Time Is Not Circular";[3] a section of Chapter 12 was included in a themed edition of *Philosophical Writings* under the title "Aristotelian Temporal Passage";[4] and some of the critiques of Ursula Coope's interpretation throughout appeared in *Mind* as part of my review of her *Time for Aristotle*.[5]

The members of the Aristotelian Society of Boise were tremendously good sports in working through my manuscript in progress and providing helpful suggestions. I've benefited immensely from their feedback and friendship.

Hilary Gaskin and Joanna Garbutt at Cambridge University Press have both been delightful to work with. Among Hilary's other editorial duties, she secured two outstanding readers whose anonymous comments on the manuscript prompted me to rethink and sometimes revise various parts of the book. I'm quite certain that my revisions fall short of their suggestions, but I am no less grateful to them for that fact. Emma Wildsmith, my project manager, was tremendously helpful in assembling the index. Christopher Feeney served as my copy editor, and his efficiency was matched only by his thoroughness.

Finally, it would have been impossible to complete this book without the loving support of my family, especially that of my wife Gayle. It takes a rather special person to put up with a philosopher at home, and Gayle is certainly that. My children, Keenan and Elizabeth, also deserve my thanks for their patience and good character. Since this is my first book, it seems appropriate to dedicate it to my first child.

This book is for you, Keenan. And it's about time.

[2] Roark (2004). [3] Roark (2003). [4] Roark (2005). [5] Roark (2009).

Figures

Abbreviations

ARISTOTLE

An.	*On the Soul* (*De Anima*)
An.Post.	*Posterior Analytics*
An.Pr.	*Prior Analytics*
Cael.	*On the Heavens* (*De Caelo*)
Cat.	*Categories*
GA	*Generation of Animals*
GC	*On Generation and Corruption*
HA	*History of Animals*
Insomn.	*On Dreams* (*De Insomnis*)
Int.	*On Interpretation* (*De Interpretatione*)
MA	*Movement of Animals* (*De Motu Animalium*)
Mem.	*On Memory* (*De Memoria*)
Met.	*Metaphysics*
Meteor.	*Meteorology*
NE	*Nicomachean Ethics*
PA	*Parts of Animals*
Phys.	*Physics*
SE	*Sophistical Refutations* (*Sophistici Elenchi*)
Sens.	*Sense and Sensibilia* (*De Sensu et Sensibili*)
Somn.	*On Sleep and Waking* (*De Somno et Vigilia*)
Top.	*Topics*

PLATO

Cra.	*Cratylus*
Epin.	*Epinomis*
Phd.	*Phaedo*
Phil.	*Philebus*

Prm.	*Parmenides*
Rep.	*Republic*
Sph.	*Sophist*
Tht.	*Theaetetus*
Ti.	*Timaeus*

Translations of *Physics* IV.10–14 are my own. The remaining translations of Aristotle are adapted from those collected in Barnes (1984). Translations of Plato are adapted from those collected in Cooper (1997), except where noted. Where a particular translation of either philosopher is problematic or otherwise interesting, I credit the translator in a footnote. Translators of other figures are credited in footnotes.

Introduction

Aristotle conceives of time as a variety of hylomorphic compound. By "hylomorphic" I mean the kind of analysis Aristotle employs at very many places in his works, according to which the thing under consideration is to be understood as a combination of matter (*hulê*) and form, or shape (*morphê*). On the hylomorphic interpretation I endorse, motion is the matter of time, and perception is its form. Aristotle defines time as "a number of motion with respect to the before and after" (*Phys.* 219b1–2),[1] by which he intends to denote motion's susceptibility to division into undetached parts of arbitrary length, a property that it possesses both by virtue of its intrinsic nature and also by virtue of the capacities and activities of percipient souls. Motion is intrinsically indeterminate, but perceptually determinable, with respect to its length. Acts of perception function as determiners; the result is determinate units of kinetic length, which is precisely what a temporal unit is.

It would be one thing to employ the conceptual framework of hylomorphic analysis as an interpretative apparatus or strategy, but I don't intend to use hylomorphism that way. I am convinced that the proper way of understanding Aristotle's view of time requires thinking of it as a variety of hylomorphic compound, because that is precisely how he himself understood it.

I take it that this view might seem quite implausible at first blush. For surely hylomorphism is most obviously suited to accounting for the nature of concrete objects like statues, houses, and animals. Conceiving of a statue as possessing two metaphysically distinct components (bronze and the shape of Hermes, for example) provides the basis for fruitful philosophical analyses along several different lines: causal properties, persistence conditions, semantic relations, and epistemic status, to name just a few.

[1] τοῦτο γάρ ἐστιν ὁ χρόνος, ἀριθμὸς κινήσεως κατὰ τὸ πρότερον καὶ ὕστερον.

But why ought we to think that time, which is at once both metaphysically fundamental and seemingly abstract, should yield to such an analysis?

The question itself reveals an unhelpful prejudice that I shall attempt to dispel in Chapter 1. But prejudice aside, it should come as no great surprise if such an approach turns out to be exegetically fruitful. Aristotle's temporal theory is developed within the *Physics*, a treatise in which he examines the objects and phenomena of nature. In the second book of the *Physics*, Aristotle states that "we think we know something only when we find the reason why it is so, i.e., when we find its primary cause" (194b18–20).[2] But "cause" (*aitia*) is ambiguous between several different senses, including material and formal causes.[3] Thus, the most complete account of any existent will include a specification of its matter and its form.[4] It would be quite surprising indeed if this general principle didn't apply to a feature of nature such as time, since the principle is alleged to apply even to linguistic objects such as syllables and arguments (195a15–17).[5] But this is not the place to argue for my thesis. I simply request that the reader consider carefully the arguments I offer in the chapters that follow. I anticipate residual resistance on points of detail that are left unfinished at the end of the book; however, if my arguments render the hylomorphic interpretation an attractive alternative to its predecessors, I will be happy to engage in debate over the details.

Now one might be inclined to think that the existence of time is a necessary condition for the existence of material objects. It seems necessary for the variety of objects that we're accustomed to interacting with, anyway. Indeed, it is quite difficult to imagine a world furnished by the kinds of things we're familiar with in which there is no time, since these sorts of things are subject to change, which, again, one might take to be a feature of the world that requires the existence of time. A world without time (according to this view) is a world without change, one populated with people who never age, fruit that never ripens, rivers that don't run, and so on. Time, then, might be thought to be an aspect of nature without which change and changing objects are impossible.

[2] εἰδέναι δὲ οὐ πρότερον οἰόμεθα ἕκαστον πρὶν ἂν λάβωμεν τὸ διὰ τί περὶ ἕκαστον (τοῦτο δ' ἐστὶ τὸ λαβεῖν τὴν πρώτην αἰτίαν).

[3] See *Phys.* II.3.

[4] Almost anything. Aristotle acknowledges the exceptions of unformed elemental matter and un-enmattered divine forms. But these are not topics of investigation for the physicist.

[5] ἅπαντα δὲ τὰ νῦν εἰρημένα αἴτια εἰς τέτταρας πίπτει τρόπους τοὺς φανερωτάτους. τὰ μὲν γὰρ στοιχεῖα τῶν συλλαβῶν καὶ ἡ ὕλη τῶν σκευαστῶν. Cf. *Met.* IV.2.

Whether this characterization of the contemporary popular view of time is correct (supposing that such a thing exists), I can't say. What is clear, though, is that Aristotle turns this picture on its head: according to Aristotle, time owes its existence immediately to motion and perception, and ultimately to material objects, substances. His motivation for endorsing such an account seems to be his commitment to the view that the strictest ontology is one which includes only instances of natural kinds (e.g. individual men, horses, trees). In such an ontology, there is no room for an item of such abstract character as time. But surely time exists *in some way or other*. And so Aristotle concludes that time must somehow derive its existence from substances. Filling out the account of this parasitic existence is the challenge that Aristotle attempts to meet in *Physics* IV.10–14, and the purpose of this work is to spell out Aristotle's answer to that challenge.

Before I set about examining the relevant texts and developing my interpretation, though, I ought to say something about why I've written this book and how it is structured.

To the first point, I feel no embarrassment in saying that this book combines two of the most philosophically exciting topics that there are: the nature of time and the philosophy of Aristotle. While many philosophers would challenge this evaluation, I take it that no one reading this Introduction is likely to lead that charge.

The philosophy of time is home to some of the knottiest conceptual problems around. Augustine is famous for confessing his own inability to articulate time's nature: "What then is time? If no one asks of me, I know; if I wish to explain to him who asks, I know not" (*Conf.* 11.14; Augustine [1980]).[6] Closer to our own day, C. D. Broad has expressed similar frustration:

> I am well aware how easy it is to talk nonsense about Time, and to mistake for arguments what are in fact merely verbal tangles. I think it is quite possible that I may have done this. I have altered my mind too often on this most perplexing subject to feel any confidence that my present opinions are either correct or well-founded.[7]

Judging from the number of academic monographs produced on the subject within the past twenty years, the philosophy of time is enjoying a good deal of attention. The explanation for this popularity is not far to

[6] quid est ergo tempus? si nemo ex me quaerat, scio; si quaerenti explicare velim, nescio.
[7] Broad (1967), 138–9.

seek: time remains a perplexing subject, and the business of philosophy is to clarify that which perplexes.

I have written about Aristotle's theory of time elsewhere, and parts of this book draw directly on those published works. But it has always seemed to me that Aristotle's view of time is sufficiently complex in itself, and so thoroughly bound up with different theories and doctrines developed elsewhere in the corpus, that a proper treatment of the account requires a book-length effort. I am glad that I have made the effort to produce a systematic reconstruction of his account of time, because in the course of doing so I have come to appreciate other, seemingly unrelated aspects of Aristotle's thinking to a far greater extent than I could have imagined.

Aristotle is a philosophical optimist, and that optimism is evident in the fact that he articulates a concise definition of time in the *Physics*. But although he is ambitious, he is not cavalier. Before he articulates his definition, he acknowledges several puzzles surrounding the nature of time, puzzles whose solution must somehow fall out of his own account. Aristotle never explicitly addresses the puzzles after giving them voice, but I shall show in Chapter 13 what his solutions must be, given the view he develops in those final five chapters in Book IV of the *Physics*. For the most part, the solutions turn out to be fairly trivial. This is at once both disappointing and gratifying: disappointing, because one might have expected fancier footwork in dealing with the puzzles; gratifying, because their triviality explains why Aristotle doesn't bother to lay them out for us, and this fact confers additional credence upon the hylomorphic interpretation.

Understanding Aristotle's definition of time is not easy. But those who are familiar with his writings know that there is little in Aristotle that is easy. His views are individually sophisticated and collectively reticulate. One can hardly work through any particular area of his thought without feeling compelled to turn from treatise to treatise with the hope that doing so might help to discover the underlying theoretical machinery. What at first seems to be a simple question of his stance on a narrow question quite often turns into an exegetical labyrinth. Such is the case with his treatment of time.

While I expect some readers to doubt that I have produced the uniquely correct reconstruction of Aristotle's temporal theory, I think it would be an error to deny that I have identified at least one path through the labyrinth. Indeed, my complaint with so many of the competing reconstructions of Aristotle's temporal theory is that they simply *couldn't* be adequate, because they fail even to countenance some of the philosophical territory that seems so obviously to stand between the opening moves and the conclusion, let alone to navigate it successfully.

The territory I have particularly in mind is Aristotle's account of perception. One recent work examining Aristotle's temporal theory – Ursula Coope's *Time for Aristotle*[8] – evidences an impressive knowledge of Aristotle's doctrines and displays enviable ingenuity. Any reader of Coope's book is bound to learn a great deal about the Aristotelian conception of time. I certainly have. However, I am convinced that neither Coope nor any of the many other interpreters of *Physics* IV.10–14 has paid sufficient attention to the central role of perception within Aristotle's theory of time. Indeed, if my own hylomorphic interpretation is correct, *all* of the previous attempts to reconstruct Aristotle's temporal theory have neglected a critical component of time's essential nature: its form.

It would be both inefficient and inappropriate for me to undertake a systematic critique of Coope's work here. It would also be inexcusable if I were not to engage her interpretation on central points of contention. Therefore, I shall quite selectively present and critically discuss her views and those of many different scholars, from ancient times to our own, as suits my purposes. I suspect that few of these philosophers would concede dialectical victory on the question at issue, and many would rightly criticize me for overlooking some part of their own reconstruction that they regard as particularly important. But my aim is not to discredit my exegetical competitors by direct refutation, for that would be a practically impossible task. I prefer to believe that my interpretation is strong enough to stand on its own legs; I engage the views of others primarily to indicate what I take its legs to be.

My strategy for defending a hylomorphic interpretation of Aristotle's account of time proceeds in four phases, each comprising a major division of this book.

In Part I I prime the pump, so to speak, by examining the background conception of time against which Aristotle developed his own view. Aristotle is a conservative philosopher in the sense that he seeks everywhere to preserve the judgments of common sense to the greatest extent possible and to integrate the insights of his predecessors.[9] Consequently, we shall do well to examine briefly the extant evidence that is relevant to determining what a typical fourth-century Athenian citizen's notion

[8] Coope (2005).

[9] For the definitive discussion of this aspect of Aristotle's methodology, see Owen (1975b).

of time might have been like. The central sources of evidence here are Hesiod, Herodotus, and Thucydides, but we shall also have occasion to touch briefly upon the time-keeping technology of Aristotle's time.

Perhaps more importantly, a well-grounded reconstruction of Aristotle's view on time (or any other topic, for that matter) must begin by examining its Platonic counterpart. For whether Aristotle is rejecting the extravagance of Plato's forms or attempting to integrate what he regards as the less radical elements of his doctrines, Aristotle's thought always begins by countenancing the genius of his mentor. The most well developed discussion of time in the Platonic corpus is to be found in the *Timaeus*, which presents its own exegetical challenges. An explication of the view presented there constitutes the bulk of Chapter 2.

Before any of this historical work can be done properly, however, we must first reflect on our own concept of time. It has become absolutely clear to me that one of the primary obstacles to understanding Aristotle's temporal theory is to unreflectively import into his view (and most particularly, into his definition of time) modern notions that are quite inappropriate within that context. We will be far less likely to fall prey to this mistake if we make a conscious effort to articulate the various facets of the concept of time that figures in our own thought on the subject. To this end, I shall make use of the theoretical apparatus developed by the twentieth-century British idealist J. M. E. McTaggart. McTaggart formulated an argument for the irreality of time that has driven some philosophers to fits. As it turns out, not only is the McTaggart-type architecture useful in preparing us for an investigation of ancient notions of time, it also serves as an interesting test for Aristotle's hylomorphic temporal theory. Toward the end of the book (specifically, in Chapter 12), I shall examine what I take to be one of the virtues of Aristotle's view, namely its immunity to arguments like McTaggart's.

Chapter 1 closes with some remarks about hylomorphism and how it figures in Aristotle's larger philosophical agenda. Appreciating the motivation behind his implementation of hylomorphic analyses will greatly aid our understanding of how his analysis of time is supposed to run.

Part II of the book is dedicated to developing the material side of the hylomorphic analysis of time. One of Aristotle's first conclusions about the nature of time in *Physics* IV.11 is that it must be some aspect of motion (219a9–10).[10] My reconstruction of the argument to this conclusion requires us to understand Aristotle as endorsing the idea that time is what

[10] ἀνάγκη τῆς κινήσεώς τι εἶναι αὐτόν.

I call an "evident proper feature" of motion. This takes us directly to the very common objection that Aristotle's definition of time as "a number of motion with respect to the before and after" is patently circular. Both motion and the *prior–posterior* relation (so the objection runs) presuppose the existence of time, so neither can very well figure in its definition. The objection is based in part on the unhelpful prejudice I address in Chapter 1, but more specifically it fails to take seriously Aristotle's analysis of motion in Book III of the *Physics*, which is developed in non-temporal terms. To deflect the objection, in Chapter 4 I examine Aristotle's definition of motion in detail and offer a novel interpretation of it. Once his view of motion is properly understood, the claim that his definition of time is objectionably circular on account of the fact that "motion" (*kinêsis*) appears in the definiens becomes implausible.

Not only does a proper understanding of Aristotelian motion vanquish the first charge of circularity, it also provides the necessary theoretical resources to address the second charge, namely the claim that the appearance of "before and after" in his definition illicitly invokes time. Appealing to the interpretation I develop of Aristotle's definition of motion, I endeavor to show that: (a) there is a distinctly kinetic (and therefore non-temporal) sense of "before and after" at work in his definition of time; (b) "the before and after in motion" denotes the material component of "the now," or the instantaneous present; and (c) the kinetic entities so denoted are intrinsically ordered and thereby provide the basis for temporal order. This project comprises Chapter 5.

Once the material constituent of time is properly understood, we may proceed to investigate the form of time, which is the focus of Part III of the book. Chapter 6 begins with an argument to the conclusion that the relevant sense of "number" in Aristotle's definition of time requires us to regard perception as time's form generally, and more narrowly of the form of "the now." This conclusion gives rise to an objection about the phenomenal character of perception, an objection whose reply is to be found in Aristotle's sophisticated theory of perception. By examining the central texts in which this theory is developed (principally the *De Anima* and the *Parva Naturalia*), we shall see that *phantasia* ("imagination") plays a crucial role in perception, as Aristotle understands it, and therefore also in his account of time. Briefly, *phantasia* serves as the basis for both memory and anticipation, thereby making possible the possession of mental states about the past and the future. *Phantasia* is also directly implicated in the perception of what Aristotle calls the "common perceptibles," among which motion is included. Clearly, Aristotle's views concerning our perception of the common perceptibles must contribute substantially to the hylomorphic analysis of his temporal theory.

With the formal analysis of time in place, we will be prepared in Part IV to address several difficult questions that remain concerning Aristotle's views on time. In particular, we shall have to determine how the hylomorphic interpretation accounts for simultaneity and other temporal relations. Celestial motion is important here, and so we shall examine briefly some of Aristotle's remarks on this subject, principally in the *De Caelo*. The nature of "the now" and temporal passage is taken up in Chapter 12. Recent scholarship challenges Aristotle's position on temporal passage, or the apparent "flow" of time, and I show how the hylomorphic interpretation produces a view that is largely immune to contemporary arguments against the possibility of temporal passage. I close the book with a few brief, speculative remarks about how the success of Aristotle's view on this issue might be emulated by contemporary theorists working in the philosophy of time.

I conclude this Introduction with a note about the conventions I've adopted in textual citation. Ever hopeful that Aristotle's views on time hold some interest for philosophers whose training is not primarily in classical philosophy, I have chosen to make the body of this book Greek-free – only transliterations will appear whenever Aristotle's own words are recounted. I put the original text in footnotes for those with Greek and also for those without but who nonetheless find it beautiful to look at.

PART I

Times new and old

CHAPTER I

McTaggart's systems

One needn't be a metaphysician to have been gripped by puzzles concerning the nature of time. Does it have a beginning? Will it ever come to an end? Why is it that we can remember the past but can only anticipate the future? What happens to objects and events of the past – do they exist in any sense at all? Are any events of the future already determined? Am I identical to any person who existed in the past? How is it that time *passes*? What *is* time, anyway? Growing up on our family farm, I would sometimes lie awake nights pondering questions such as these, unsure how I might even begin to find their answers.

As C. D. Broad's remark quoted in the Introduction suggests, much of the difficulty we experience in discussing and contemplating time arises from the language in which we frame the propositions of speech and thought. Certain facts about the English language (e.g. that temporal adverbs are frequently vague or are used ambiguously, that verbs tokened in writing and speech indicate grammatical tense, or even that the word "time" can function as the subject of a sentence) give rise to challenges for those who would inquire into the nature of time.

But these are no less facts about Attic Greek than they are about English, and Aristotle – arguably the first philosopher to explicitly attempt to unravel the snares inherent in language – must have faced the same sort of challenges that arise from them when he composed his own theory of time, an account that he develops in chapters 10–14 of Book IV of the *Physics*. Indeed, he goes to some trouble in IV.13 to consider the nuances of the expressions "at some time" (*to pote*), "recently" (*to arti*), "immediately" (*to êdê*), "long ago" (*to palai*), and "suddenly" (*to exaiphnês*) (222a25–b30). None of this, of course, is meant to suggest that Aristotle's theory of time should be familiar or even particularly accessible to us today.

Aristotle's philosophical successes are indebted, at least in part, to his insistence on taking stock of the standing opinions regarding his subject before launching his own examination. By doing so, he seems to believe that he is better situated to avoid the mistakes of his predecessors and to capitalize on their insights. In this regard, his approach is both modest and conservative.

If we hope to properly understand Aristotle's account of time, we would do well to emulate his method. As interpreters of his view, we are obliged to examine *two* sets of standing opinions: those that provided the context for Aristotle's thinking about the subject of time, and those that influence our own thoughts on the subject. While it is fairly obvious that any responsible treatment of Aristotle's account must address the former, it is far less so that the latter would also figure in such a treatment. But in fact it is absolutely critical that we have a clear view of our own conceptual position before looking into Aristotle's, and yet interpreters have been too eager to examine Aristotle's account without first giving some thought to their own concept of time. Proceeding in this way, they are all too liable to fall victim to certain prejudices and confusions that we must try to avoid.

One of the easiest mistakes to make in thinking about time is to suppose (at least tacitly) that our concept of time is monolithic. It most certainly is not. Rather, it is a family of concepts – difficult to articulate in themselves – that are related in ways that are difficult to discern. It will be worthwhile, then, to make some attempt to tease apart the several concepts that form the fabric of our concept of time, for it is very unlikely that we will succeed in making sense of Aristotle's conception of time if we are insensitive to our own.

Probably the best-known analysis of the concept of time was offered in the early twentieth century by J. M. E. McTaggart, who claims that there are two fundamentally different ways of conceiving of time – one he calls the "*B* series," and the other he calls the "*A* series."[1] Each series is a metaphysically primitive structure consisting of an ordered, linear continuum of events. The difference between the two series is the way in which their members are ordered.

The elements of the *B* series are ordered by the relations *earlier than* and *later than*, which are relations that hold sempiternally: if it is ever true that event e_1 is earlier than event e_2, then it is always true that e_1 is earlier than e_2. As such, these temporal relations are temporally stable. Elvis Presley's death will always have occurred later than the Apollo 11 landing, and the fall of Rome will always have occurred earlier than the collapse of the World Trade Center towers. This is a way of conceiving of

[1] McTaggart (1927), ch. 3.

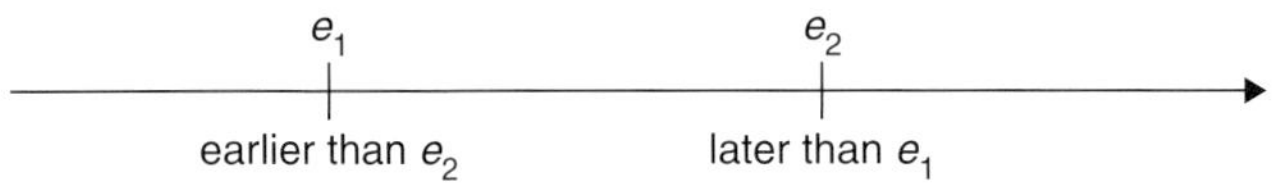

Figure 1. A model of McTaggart's *B* series.

time as a static feature of the cosmos, a feature that might fairly be called "temporal extension" or "duration." Inherent in this conception are other essential features, such as directionality, ordering relations, and simultaneity. This conception of duration – the notion that time is (or that events in time are) "stretched out" – can be represented graphically by means of a simple diagram.

Figure 1 is a model of a temporal interval of indefinite duration, which consists in a continuum of events standing in relative proximity to one another (represented by the line) and ordered by the asymmetric, temporally stable relations of *earlier than* and *later than* (represented by the arrowhead, such that event e_1 is earlier than e_2 iff "e_1" is located to the left of "e_2," later than e_2 iff "e_1" is located to the right of "e_2," and simultaneous if they are co-located). This is the static aspect of our concept of time that McTaggart meant to capture in his *B* series.

Most folks do not take time to be such a static sort of thing, and McTaggart himself argues that our concept of time cannot simply be the concept of the *B* series, since the latter, he claims, does not provide what he takes to be the necessary conditions for genuine change, and he thinks that no defender of time's existence will deny that there is genuine change.[2]

The *A* series is an extension of the *B* series in the sense that it includes all of its features but adds another. Whereas the elements of the *B* series are ordered only by the relations of *earlier than* and *later than*, the elements of the *A* series are further ordered by the additional relations *past*, *present*, and *future*.

The relata standing in these relations are the same elements that constitute the *B* series, but specifically insofar as each of them enjoys a certain status in the series. The distinguishing features of the *A* series are the facts that (a) whenever one cares to inspect or describe the series, *exactly one* of its elements temporarily enjoys the privileged status of being absolutely

[2] McTaggart (1927), §§310, 311. For an extended argument that the *B* series is an accurate model of reality, see Mellor (1981). To my mind, Savitt (2001) shows once and for all that McTaggart's argument fails because it rests on an ineliminable equivocation.

present;[3] and (b) *every* element enjoys this status at some time or other. The central idea of the *A* series is that there is a unique and ever-changing absolute present, relative to which every other element of the series is either absolutely past or absolutely future.

It is impossible to represent the dynamic aspect of temporal passage by means of a printed representation, since the very essence of passage is the fact that something *changes* among the elements of the series – namely, which of the elements is absolutely present. Still, it is possible to represent temporal passage by means of a model that is understood to be a kind of "snapshot." Where "N" indicates which element of the series is absolutely present (*now*), Figure 2 is just one model of the temporal continuum consisting in duration and passage:

Figure 2. A model of McTaggart's *A* series.

What one must understand when one sees that the absolute present is located at a particular point along the line is that its status (ignoring continuity issues) is analogous to that of a single frame of film in an operating movie projector: every point on the line is absolutely present at some time or other, just as every frame eventually passes between the light and lens of the projector. This is a way of conceiving of time as being essentially dynamic, and it is the way that most of us quite unreflectively think about time. While it makes no sense to say that "time marches on" or to ask "where has the time gone?" within the *B* series, such sayings are quite at home within the *A* series, as passage is its central feature. Indeed, McTaggart seems to think that passage is the essential feature of time itself.

McTaggart maintains that there is a contradiction inherent in the notion of temporal passage (which is what ultimately leads him to deny the reality of time), but I don't wish to examine his argument here.[4] The relevant matter for our present interests is the fact that McTaggart has

[3] More properly, the absolute present is a cut across simultaneous events, not a single event – see Chapter 11.

[4] In Chapter 12 I shall present a version of McTaggart's argument developed by Dummett (1978) and show that, even if the argument is effective against other conceptions of temporal passage, Aristotle's conception of passage is immune to it.

Figure 3. A model of McTaggart's *A* series with periodicity.

identified several distinct concepts that jointly form our concept of time in general: temporal extension, with its related concepts of directionality, relative order, and simultaneity; and temporal passage, with its related concepts of the past, present, and future.

Probably because it was not required for his argument, McTaggart neglects an additional concept inherent in our notion of time – namely, periodicity. This concept accounts for the fact that one finds regularly recurring patterns in time. That is, in spite of the fact that time is continuous, it is also episodic, punctuated by divisions that mark out more or less natural segments of time (e.g. days, seasons, and years).

Now the concept of periodicity seems to be less basic than extension and passage, but it is not entirely clear to which of them it is subservient. There is some reason for thinking that it belongs to both of them. That there are regularly recurring patterns, or periods, in time is possible only in virtue of certain features of temporal extension; however, periodicity is as remarkable as it is to us because we experience it serially. Periodicity without serial experience amounts to nothing more than a geometrically regular structure, and the periodic nature of time seems to be more than just this. Still, it is helpful to have an image in hand that can be used to contrast periodicity with the other two concepts. Adopting fairly obvious conventions, the diagram in Figure 3 will serve to help us visualize a portion of the temporal continuum informed by the three central concepts that form the core of our concept of time in general: duration, passage, and periodicity.

While the *A* series incorporates the temporal extension of the *B* series and is able to distinguish between past and future events, it is not so structured as to *quantify* temporal displacement. So while it is true that, as they stand in the *A* series, e_1 and e_2 are respectively past and future, one cannot determine how far into the past or future they lie, or even how far they are from one another. With the incorporation of periodicity into the *A* series, we may say that e_1 took place last Monday, that e_2 will take place next Monday, and (thus) that there is a fortnight separating the two events.

It is difficult to overstate the importance of being sensitive to these distinct elements present in our concept of time. If one begins one's investigation by supposing that Aristotle takes time to consist in some determinate combination of them, one will likely find oneself facing insuperable obstacles. And if – as it turns out, I think – Aristotle's conception doesn't match our own, these distinctions will prove indispensable in articulating the differences between the two.

CHAPTER 2

Countenancing the doxai

Having distinguished the various aspects of our own concept of time, we should now examine what I take to be for Aristotle two very obvious though importantly different sources of thought on the subject of time: the common conception of time in fourth-century Athens, and his philosophical predecessors' views on time.

While Aristotle does not explicitly appeal to the first source of thought in the course of developing his theory, he was not immune to the sway of "common wisdom" in his day.[1] Certain features of his account clearly engage some of the remarks on time made by antiquity's greatest historians, and as we shall see, there can be no doubt that his view is driven in part by considerations having to do with the two types of chronometers with which he would have been very familiar.

As the second source of thought, the views of Aristotle's predecessors are always given consideration in his own work, and Plato figures as a looming presence throughout all of Aristotle's treatises. It is my view that Aristotle's theory of time is profoundly influenced by Plato, and much insight into Aristotle's project can be gained when it is cast in the light of Plato's theory. Although there are a number of dialogues containing considerations relevant to a discussion of time, Plato's most developed thoughts on the subject are contained in the *Timaeus*.[2] Consequently, I shall focus my attention on this work in particular.

[1] See *Phys.* 223b24–4a2, for example.

[2] Outside of the *Ti.*, Plato discusses temporal passage (and related issues, such as change) in the *Rep.*, *Prm.*, *Laws*, and *Cra.*; he discusses the relation between truth and time in the *Rep.* and *Cra.*; he discusses eternity in the *Rep.*, *Phd.*, and *Epin.*; he also offers a brief remark at *Phil.* 30c that echoes/anticipates his account in the *Ti.* according to which the demiurge is responsible for the existence of time.

2.1 OLD TIME AND ANCIENT CHRONOMETERS

Even to the most primitive of peoples, it must be plainly evident that the activities of human beings are governed by time. We have only finite life spans, and one's conception of oneself tracks the growth, maturation, and ultimate degeneration of the physical body and its capacities. Even at a more mundane level, our daily activities are regulated in time both by internal mechanisms (including the roughly twenty-five-hour "circadian rhythm" of cyclical metabolic activity and sleep governed by hormone production) and by external factors (the number of hours of sunlight, for example). One needn't have a very sophisticated understanding of these mechanisms to be aware of the fact that one is subject to them. Just waking up to the rising sun and feeling the stiffness of age is evidence enough.

This awareness of "natural time," to give it a name, seems to have had two importantly different aspects in the ancient world. The first aspect was one of awe and wonder, regarding time as a fundamental constituent of the cosmos as well as a source of being. Orphic verses, Homer's *Iliad*, and Hesiod's *Theogony* all describe how Gaia (earth) and Ouranos (heaven) begot Chronos (time), a tradition surely inherited from Babylonian and other Near Eastern cultures before the first millennium.[3] Chronos is portrayed in this tradition as a multi-headed snake with wings, suggesting to some scholars indebtedness to Iranian theology.[4] This mythological aspect of the ancients' conception of time is interesting and important from a historical point of view, but it is of relatively little instrumental value in the present context. It is worth noting, at least in part, because Plato acknowledges the mythological tradition in the *Cratylus* and the *Timaeus*, though his own account of the creation of time is fundamentally different.

The second and more germane aspect of the ancient awareness of natural time was essentially practically oriented, since agricultural, religious, and maritime activities were organized around seasonal changes.[5] Additionally, many of Greece's prominent cultural events were held according to a regular schedule. In particular, organizing the Panhellenic games required a reliable method of dating years. The four sets of games were staggered in such a way as to permit an athlete to compete in each

[3] Aristotle acknowledges the influence of Babylonian and Egyptian astronomy in *Cael.* II.12.

[4] Kirk and Raven (1957), 39.

[5] See, e.g., Hesiod (1964), 609–17, 641–5.

of them. The Olympic games marked the beginning of a four-year cycle (called an "Olympiad," the first of which is dated to 776 BCE), with the Nemean and Isthmian games each held in both the second and fourth year, and the Pythian games in the third year.

Of course, the Greeks did employ an annual calendar, and Herodotus tells us that its twelve months were named after a tradition inherited from the Egyptians.[6] In the *History*, Herodotus indicates that the standard year in sixth-century Greece was 336 days in length, with twelve months of twenty-eight days each.[7] Every other year, an intercalary month thirty days long was added to calibrate (not perfectly but more nearly) the legal calendar with the solar year. The dating system is impressive in its complexity, but it is important to note that there is no provision for the numbering of years, owing to the lack of a universally recognized epoch or origin. The Gregorian calendar in use today treats the birth of Christ as the epoch,[8] whereas the Julian calendar, implemented by Julius Caesar in 46 BCE, tracked years by the name of the officiating consul, each of whom assumed office on January 1. The first Olympiad was the most epoch-like device used in Aristotle's day (e.g. "the thirtieth Olympiad" would designate a period spanning the years 656–653 BCE, inclusive), but no convention existed for designating individual years. It was not until 525 CE that the Roman abbot Dionysius Exiguus lobbied for the adoption of the *Anno Domini* system employed in the modern Gregorian calendar.

The lack of a common reference device for years prompts Herodotus to denote them in the *History* by appealing to significant events proximate in time and indicating the lapse of time in the interval: "in the eleventh year of their exile" (21); "in the course of the fifth month from the time of the departure of his messengers" (29); "about the time of Cambyses' last sickness the following events happened" (190).[9] Thucydides, another historian of antiquity renowned for his commitment to precision (if not accuracy), employs similar devices in dating events in the Peloponnesian War, as "in the summer of the third year of this war," and the like. The way in which dating practices such as these are reflected in Aristotle's account of time will be discussed in Chapter 6.

Despite its various inadequacies, the Greek calendar represents a sophisticated grasp of measuring long periods of time. The basis for the

[6] Herodotus (1947), 82. [7] *History* 1.32.

[8] The adoption of an epoch is welcome, but this particular choice is problematic for two reasons: it lacks a year zero, and there is widespread scholarly agreement that Jesus was in fact born several years before 1 CE.

[9] Herodotus (1947).

calendar's structure, including its imposition of intercalary months, was a device used primarily to record the sun's angle in the sky: the sundial (*gnomon*). Fourth-century Athens witnessed the proliferation of such devices used to mark out briefer periods of time, and any citizen of that period in time would also have been familiar with another type of mechanical chronometer: the water-clock (*klepshydra*, so called because the vessels from which they were fashioned were initially used to "steal" water from cisterns).

Sundials have enjoyed centuries of popularity as garden decorations, and when they are properly designed and installed, they are sometimes even treated as reliable clocks. The first sundials were not designed to divide the solar day into sub-periods, however, but rather to mark the passing of the seasons. Their design probably allowed an observer to note the relative angle of the sun by means of a stylus or bar that cast a shadow on a graduated tablet. There are reports that Anaximander designed and installed a device of this type in Sparta.[10]

The Egyptians attempted to divide the solar day into sub-periods as early as 1400 BCE, but their sundials could not have served as reliable clocks (in the sense that they could not facilitate the comparison of relative lengths of distinct sub-periods), because it was not until *c.* 350 BCE that Eudoxos of Knidos invented a serviceable theory of gnomonics, which prescribes techniques of compensating for the seasonal change in solar angle.[11] Despite this, the Greeks fabricated "day-dials" at least a full century before Eudoxos' theory, and some of them were put to official use. The astronomer Meton reportedly installed a day-dial on the Pnyx (a hill just west of the Acropolis) in 433 BCE, probably to mark the meeting times of the Assembly.[12] This pre-Eudoxian dial would have served its purpose regardless of its reliability (in the sense mentioned above), since it would have provided a readily visible indicator by which an assembly member could have estimated, practically at any time, how long he had before he and his fellows were to assemble – or whether he was late, for that matter. Other devices (such as the sounding of a bell) would not have afforded this degree of freedom and efficiency.

Though he does not mention sundials in any extant work, Aristotle would have been familiar with them. There is, however, explicit reference in the corpus to water-clocks. Similar to the hourglass in principle and

[10] Diels (1912), ch. 2, fragment 4: οὗτος πρῶτος γνώμονας κατεσκεύασε πρὸς διάγνωσιν τροπῶν τε ἡλίου καὶ χρόνων καὶ ὡρῶν καί ἰσημερίας.

[11] Janich (1985), 167–8. [12] Brumbaugh (1966), 71.

design, the earliest of these devices were nothing more than vessels with a small hole at the base that could be uncovered to release the vessel's liquid contents. The Greeks probably inherited this simple chronometer from the Egyptians, who employed them in the courts to ensure that the parties involved were allowed equal time to plead their cases.[13]

Very near the end of the *Constitution of Athens*, Aristotle mentions the use of water-clocks as he describes the elaborate rules governing the timing of cases heard in the fourth-century courts:

> Water-clocks are provided, having small supply-tubes, into which the water is poured by which the length of the pleadings is regulated. Ten gallons [*choes*; one *chous* ≅ three liters] are allowed for a case in which an amount of more than five thousand drachmas is involved, and three for the second speech on each side. When the amount is between one and five thousand drachmas, seven gallons are allowed for the first speech and two for the second; when it is less than one thousand, five and two. Six gallons are allowed for arbitrations between rival claimants, in which there is no second speech. The official chosen by lot to superintend the water-clock places his hand on the supply-tube whenever the clerk is about to read a resolution or law or affidavit or treaty. When, however, a case is conducted according to a set measurement of the day, he does not stop the supply, but each party receives an equal allowance of water. The standard measurement is the length of the days in the month of Poseidon * * * The measured day is employed in cases when imprisonment, death, exile, loss of civil rights, or confiscation of goods is assigned as the penalty. (67.2–5)[14]

The complexity of the rules of court procedure is impressive in itself, but this passage contains another noteworthy idea. The typical time allotment for pleading is expressed in terms of liquid volumes; however, Aristotle says that some cases are "conducted according to a set measurement of the day," strongly suggesting that these sorts of cases proceeded not according to some fixed but arbitrarily determined length of time (e.g. "six gallons' worth of time"), but rather according to a more well-defined metric by which the length of a standard day could be carved up without

[13] Janich (1985), 168.

[14] εἰσὶ δὲ κλεψύδραι αὐλίσκους ἔχουσαι ἔκρους, εἰς ἃς τὸ ὕδωρ ἐγχέουσι, πρὸς ὃ δεῖ λέγειν περὶ τὰς δίκας. δίδοται δεκάχους ταῖς ὑπὲρ πεντακισχιλίας καὶ τρίχους τῷ δευτέρῳ λόγῳ, ἑπτάχους δὲ ταῖς μέχρι πεντακισχιλίων καὶ δίχους, πεντάχους δὲ ταῖς ἐντὸς καὶ δίχους, ἑξάχους δὲ ταῖς διαδικασίαις, ὕστερον λόγος οὐκ ἔστιν οὐδεις. ὁ δ' ἐφ' ὕδωρ εἰληχὼς ἐπιλαμβάνει τὸν αὐλίσκον, ἐπειδὰν μέλλῃ τινὰ ἢ νόμον ἢ μαρτυρίαν ἢ τοιοῦτόν τι ὁ γραμματεὺς ἀναγιγνώσκειν. ἐπειδὰν δὲ ᾖ πρὸς διαμεμετρημένην τὴν ἡμέραν ὁ ἀγὼν, τότε δὲ οὐκ ἐπιλαμβάνει αὐτόν, ἀλλὰ δίδοται τὸ ἴσον ὕδωρ τῷ, τε κατηγοροῦντι καὶ τῷ ἀπολογουμένῳ, διαμετρεῖται δὲ πρὸς τὰς ἡμέρας τοῦ Ποσιδεῶνος * * * διαιρεῖται δ' ἡ ἡμέρα ἐπὶ τοῖς … ἀγώνων ὅσοις πρόσεστι δεσμὸς ἢ θάνατος ἢ φυγὴ ἢ ἀτιμία ἢ δήμευσις χρημάτων, ἢ τιμῆσαι δεῖ ὅ τι χρὴ παθεῖν ἢ ἀποτεῖσαι. Some ten lines of the text are missing at the lacuna.

remainder. In his commentary on this treatise, Rhodes reports the work of German scholar B. Keil:

> [A]s the basis for the division of the day the Athenians take a day in Poseidon, which is their sixth month and will normally have contained the shortest day; an allowance which was sufficient for business to be completed on the shortest day will have been more than sufficient for other times of the year … The length of the shortest day in Athens is given as 9 hrs 28 minutes by Keil … [who] suggested that the court day was divided into twelve "winter hours" of (fractionally under) 48 minutes, represented by an amphora of water, and each "hour" was subdivided into 12 *choes* of 4 minutes.[15]

It is a fascinating fact that this kind of *klepshydra* manages to divide a fixed period of sunlight into type-repeatable units which are not themselves abstracted from any movement of the sun. Indeed, water-clocks were used to calibrate sundials prior to Eudoxos' development of gnomonics around 350 BCE.[16] We shall have opportunity to note a philosophically relevant difference between water-clocks and sundials in Chapter 11, after Aristotle's hylomorphic analysis of time has been presented.

Plato also mentions *klepshydra* in such a way as to imply their relatively regular use in public life, especially in the courts. In the *Theaetetus*, he compares the discussions of lawyers and orators to those of philosophers, claiming that participants in the former discussions are slaves to the water-clock, neither free to select their topic nor to determine how long they will speak (172d9–e1, 201a10–b4).[17] Besides corroborating the ubiquity of such devices, this provides us the opening to briefly examine Plato's account of time.

2.2 TIME IN THE *TIMAEUS*

It would be both impractical and unnecessary to attempt to develop a comprehensive interpretation of Plato's theory of time here. Instead, my plan is to sketch the elements of Plato's account of time that will be relevant to my examination of Aristotle's theory. There are two central questions to be addressed in the course of our discussion: (i) how does Plato define time? and (ii) how does the concept of time captured by the definition differ from our own concept of time, as it was discussed in Chapter 1?

[15] Rhodes (1981), 726. Rhodes cites Keil (1902).
[16] Janich (1985), 168.
[17] οἱ δὲ ἐν ἀσχολίᾳ τε ἀεὶ λέγουσι – κατεπείγει γὰρ ὕδωρ ῥέον … ἢ σὺ οἴει δεινούς τινας οὕτω διδασκάλους εἶναι, ὥστε οἷς μὴ παρεγένοντό τινες ἀποστερουμένοις χρήματα ἤ τι ἄλλο βιαζομένοις, τούτους δύνασθαι πρὸς ὕδωρ σμικρὸν διδάξαι ἱκανῶς τῶν γενομένων τὴν ἀλήθειαν. In both passages, Plato refers to the clocks as "the water."

Before taking up these questions, an interpretative point is in order. I claim that the value of understanding Plato's theory of time lies in its function as an explanatory device, by way of its influence on Aristotle's view. Underlying this claim is the supposition that Aristotle took Plato to *have* a carefully considered theory of time. This supposition is borne out by numerous remarks in the corpus. One of these that speaks quite clearly is in *Physics* VIII.1, where Aristotle identifies Plato as his only predecessor to maintain that time was created – a view advanced by the dramatic character Timaeus in his eponymous dialogue (251b17 ff.).[18] Aristotle is critical of this view, as he is of other aspects of the cosmology developed in the *Timaeus*, explicitly attributing them to Plato himself.[19] So there is rather compelling evidence that Aristotle took the *Timaeus* to express Plato's cosmology, including his theory of time.

There remain, of course, the much more difficult questions of whether the cosmological view developed in the *Timaeus* in fact represents Plato's own, and if so, whether he intends it to be taken literally or in some weaker way, as a "likely myth." Scholars are divided on these questions. A. E. Taylor, for instance, argues that Timaeus' creation account must be regarded as a brainchild of Plato that he never intended to foster. As Taylor understands the work, Plato cobbled together a view from elements of Empedoclean biology and Pythagorean mathematics.[20] The other view is best represented by F. M. Cornford, who rejects Taylor's reading, claiming that, though they certainly would have been in a position to do so if they thought it were correct, no ancient commentator (including Aristotle and the Neoplatonists) advocates the interpretation that Taylor puts forward.[21] While I find Cornford's argument persuasive, there is no need to examine it (or any of the other more challenging questions regarding the *Timaeus*) in detail here. The terms of my discussion, like those of Aristotle's, presuppose that the *Timaeus* contains a genuinely Platonic theory about the nature and creation of time and the cosmos. Consequently, I shall use Plato's and Timaeus' names more or less interchangeably, as context suggests. I invite those who favor Taylor's reading of the *Timaeus* to make such discriminations as they see fit.

[18] Cf. Vlastos (1965), 384 ff.

[19] See esp. *Cael.* 280a27. Cf. *Met.* 1071b31 ff. and *An.* 406b26 ff.

[20] Taylor (1928), 18. One bit of prima facie support for Taylor's interpretation is the caveat Timaeus gives regarding the literal truth of his account at 27c1 ff., which bears a striking resemblance to an argument against the knowability of the visible realm offered by Socrates in *Rep.* v.

[21] Cornford (1937), ix–xi.

Plato's general cosmogonic view is that of a divine craftsman (*dêmiourgos*) imposing order on raw materials standing in a chaotic condition; and as we shall see, this is true of his account of the genesis of time specifically. However, there is an initial stumbling block that stands in the way of understanding what looks to be Plato's official definition of time, because his statement of it is syntactically ambiguous. It will be worthwhile to examine the passage at length:

> When the father who had begotten [the universe] saw it set in motion and alive, a shrine brought into being for the everlasting gods, he rejoiced, and being well pleased he took thought to make it yet more like its pattern. So as that pattern is the Living Being that is forever existent, he sought to make this universe also like it, so far as might be, in that respect. Now the nature of that Living Being was eternal, and this character it was impossible to confer in full completeness on a generated thing. But he took thought to make, as it were, a moving likeness of eternity [*eikô … kinêton tina aiônos*]; and, at the same time that he ordered the Heaven, he made, of eternity that abides in unity, an everlasting likeness moving according to number – that to which we have given the name Time [*touton hon dê chronon ônomakamen*]. (37c6–d7)[22]

This is Cornford's translation. For our present purposes, the text of interest is the two phrases that flank the semicolon, where Plato articulates his definition of time. The final phrase contains Plato's first mention of time by name, so one might expect it to express his "official" definition. Unfortunately, the Greek is ambiguous insofar as the antecedent of the pronoun in the relative clause may be either "*arithmon*" ("number") or "*eikôn*" ("likeness," "image").[23] Cornford preserves this ambiguity in his translation, but not all translators do. Donald Zeyl, for example, opts to specify the former alternative in his translation: "This number, of course, is what we now call 'time'."[24] According to Zeyl's translation, Plato's conception of time bears a striking resemblance to Aristotle's, which also

[22] Ὡς δὲ κινηθὲν αὐτὸ καὶ ζῶν ἐνόσεν τῶν ἀιδίων θεῶν γεγονὸς ἄγαλμα ὁ γεννήσας πατήρ, ἠγάσθη τε καὶ εὐφρανθεὶς ἔτι δὴ μᾶλλον ὅμοιον πρὸς τὸ παράδειγμα ἐπενόησεν ἀπεργάσασθαι. καθάπερ οὖν αὐτὸ τυγχάνει ζῷον ἀίδιον ὄν, καὶ τόδε τὸ πᾶν οὕτως εἰς δύναμιν ἐπεχείρησε τοιοῦτον ἀποτελεῖν ἡ μὲν οὖν τοῦ ζῴου φύσις ἐτύγχανεν οὖσα αἰώνιος, καὶ τοῦτο μὲν δὴ τῷ γεννητῷ παντελῶς προσάπτειν οὐκ ἦν δυνατόν· εἰκὼ δ' ἐπενόει κινητόν τινα αἰῶνος ποιῆσαι, καὶ διακοσμῶν ἅμα οὐρανὸν ποιεῖ μένοντος αἰῶνος ἐν ἑνὶ κατ' ἀριθμὸν ἰοῦσαν αἰώνιον εἰκόνα, τοῦτον ὃν δὴ χρόνον ὠνομάκαμεν.

[23] εἰκών is a feminine noun, but it still may be the antecedent of τοῦτον, since it takes masculine adjectives (as is evidenced in the preceding line).

[24] Zeyl (1997). I am grateful to S. Marc Cohen for calling my attention to this translation. Though I reject Zeyl's reading of this phrase, I find his translation of the *Ti.* preferable to Cornford's (1937) in many respects and adopt it henceforth for all quotations from the work. It is interesting that, in his commentary, Cornford discusses neither the ambiguity nor what he takes the antecedent of the pronoun to be. Nonetheless, his interpretation seems to run along the lines of the one I sketch here, taking the motion of the heavenly bodies to be a sort of "celestial clock" (cf. Cornford [1937], 102).

apparently defines time as a kind of number. Now most translators who specify the antecedent of the pronoun opt for "image," since the description of time in the preceding phrase is as a "moving likeness of eternity." This ought to count as some reason to prefer "image" over "number." Additionally, Zeyl's reading renders Timaeus' discussion of time profoundly obscure at points. Consider his remarks at 38a7–8:

> These [*was* and *will be*], rather, are the forms of time that have come into being – time that imitates eternity and circles according to number.[25]

The final phrase demands that one be able to make sense of the idea of time circling according to number. Anyone who has watched television or been to the cinema (or even made shadow-puppets, or looked in a mirror, for that matter) understands perfectly well what moving images are. Such images can "circle according to number," provided that the image has as its cause an object that moves in a regular circle (say, a Ferris wheel or the hands of a clock). On the other hand, if Plato regards time as a kind of number, as Zeyl maintains, this passage presents us with moving numbers circling according to number, and it is quite difficult to make good sense of this.

Probably the most revealing passage is at 39c7–d2, where Timaeus characterizes time in perfectly unambiguous language:

> And so people are all but ignorant of the fact that time really is the wanderings of these bodies [*chronon onta tas toutôn planas*] bewilderingly numerous as they are and astonishingly variegated.[26]

As far as I can see, this passage constitutes an insurmountable obstacle to anyone who would advocate Zeyl's interpolative translation, since it states in unequivocal terms that time is the wandering movements of certain heavenly bodies.[27] There is no reason why we ought to strictly identify these movements with any number as such. On the other hand, such celestial movements could easily constitute the "moving image of eternity" mentioned in the penultimate phrase of 37c6–d7.

Therefore, there is no reason to think that Plato places time in the genus *number*. Rather, he takes time to be an image of eternity consisting

[25] ἀλλὰ χρόνου ταῦτα αἰῶνα μιμουμένου καὶ κατ' ἀριθμὸν κυκλουμένου γέγονεν εἴδη.

[26] ὥστε ὡς ἔπος εἰπεῖν οὐκ ἴσασιν χρόνον ὄντα τὰς τούτων πλάνας, πλήθει μὲν ἀμηχάνῳ χρωμένας, πεποικιλμένας δὲ θαυμαστῶς.

[27] A note on terminology: as I use the expression, "movement" denotes a particular instance or occurrence of motion, and thus represents a token-concept. In contrast, I use "motion" to denote either a type-concept which is tokened by distinct but similar movements, or, more generally, the entire class of phenomena that Aristotle takes as one of his subjects of inqiry in the *Physics* and elsewhere. I do not assume that this distinction of terms reflects standard usage – indeed, there seems to be no well established usage to reflect. Nonetheless, the distinction is quite useful, provided that it is preserved by some convention or other.

in planetary and astral motion. But what does this mean, exactly? Here it will prove helpful to take Plato's account of the genesis of the four elements as our model, recalling his general cosmogonic view.[28] In the passage below, Timaeus prefaces his presentation of the theory by describing the chaotic state of what was to become the ordered cosmos:

> Indeed, it is a fact that before [the universe came to be] the four kinds all lacked proportion and measure, and at the time the ordering of the universe was undertaken, fire, water, earth, and air initially possessed certain traces of what they are now. They were indeed in the condition one would expect thoroughly god-forsaken things to be in. So, finding them in this natural condition, the first thing the god then did was to give them their distinctive shapes, using forms and numbers [*eidesi te kai arithmois*].
>
> Here is a proposition we shall always affirm above all else: The god fashioned these four kinds to be as perfect and excellent as possible, when they were not so before. It will now be my task to explain to you what structure each of them acquired, and how each came into being. (53a7–c1)[29]

What follows these introductory remarks is Plato's ingenious account of the four elements, according to which each is a type of molecule or corpuscle – a three-dimensional regular polygon whose variously many surfaces (the "numbers" mentioned at 53b5) consist of different combinations of equilateral and scalene triangles (the "forms"), which function as Plato's atoms. The account is powerful enough to explain certain observable features of each element (e.g. the relative mobility of an elemental type as compared to the other three is explained by appeal to its shape and the size of its corpuscles), but the true ingenuity of the account becomes evident at 56c8 and following. There Plato explains how elemental transmutation is possible by invoking the rationality of the structural features of the elemental types. For instance, a single corpuscle of water (whose twenty faces each consist of six scalene triangles) can decompose into one corpuscle of fire (whose four faces each consist of six scalene triangles) and two corpuscles of air (each of whose eight faces are individually composed of six scalene triangles) without any residue (56d1–c7). Plato explains other elemental transmutations in similar, mathematically

[28] I am grateful to David Keyt for suggesting this line of thought.

[29] καὶ τὸ μὲν δὴ πρὸ τούτου πάντα ταῦτ᾽ εἶκεν ἀλόγως καί ἀμέτρως· ὅτε δ᾽ ἐπεχειρεῖτο κοσμεῖσθαι τὸ πᾶν, πῦρ πρῶτον καὶ ὕδωρ καὶ γῆν καὶ ἀέρα, ἴχνη μὲν ἔχοντα αὑτῶν ἄττα, παντάπασί γε μὴν διακείμενα ὥσπερ εἰκὸς ἔκειν ἅπαν ὅταν ἄπῇ τινος θεός, οὕτω δὴ τότε πεφυκότα ταῦτα πρῶτον διεσχηματίσατο εἴδεσί τε καὶ ἀριθμοῖς. τὸ δὲ ᾗ δυνατὸν ὡς κάλλιστα ἄριστά τε ἐξ οὐχ οὕτως ἐχόντων τὸν θεὸν αὐτὰ συνιστάναι, παρὰ πάντα ἡμῖν ὡς ἀεὶ τοῦτο λεγόμενον ὑπαρχέτω· νῦν δ᾽ οὖν τὴν διάταξιν αὐτῶν ἐπιχειρητέον ἑκάστων καὶ γένεσιν ἀήθει λόγῳ πρὸς ὑμᾶς δηλοῦν.

rigorous fashion, giving us what is very likely the first explicit articulation of a "preservation principle" of physics and the first formal chemical theory in Western history.

With Plato's account of the elements in view, we are now in a position to diagnose more precisely in what way Zeyl's translation is unsatisfactory, though in doing so we shall come to appreciate its spirit. As we have seen, Plato's account of the elements is highly mathematized; however, it would be a mistake to attribute to him the view that each of the elements *just is* a number. A corpuscle of fire is composed of four triangular faces, but it is not Plato's view that fire is identical to the number four. Rather, fire is that which *results from* the imposition of the appropriate (geometrical, mathematical) form upon the unformed elemental material mentioned at 53b2.

The same considerations apply to Plato's view of time. Time is not a number, but rather the result of imposing the appropriate structure upon unformed raw materials. In this case, though, the raw material is not ordinary matter – it is motion:

> The god wanted everything to be good and nothing to be bad so far as that was possible, and so he took over all that was visible – not at rest but in discordant and disorderly motion [*plêmmelôs kai ataktôs*] – and brought it from a state of disorder to one of order, because he believed that order was in every way better than disorder. (30a2–6)[30]

Even though there is no explicit mention of number in this passage, the idea of being orderly or prescribed (*taktos*) invites us to think quantitatively.[31] Given this fact along with the similarity of language in this passage to that of the one quoted above in which Plato discusses the genesis of the four elements, we ought to read the "according to number" qualification in Plato's definition of time as indicating that, like their elemental counterparts, time and the pre-creation disorderly motion are distinguished from one another by the presence or absence of "proportion and measure … forms and numbers" (53a–b).

Time, then, for Plato is the orderly, periodic motion of the cosmos. The cosmos is an image of the form of the Living Being itself (*zôon auto*). One aspect of this form is its eternality, and this is reflected in the image by the regularly repeating movements of the heavenly bodies: sequential years

[30] βουληθεὶς γὰρ ὁ θεὸς ἀγαθὰ μὲν πάντα, φλαῦρον δὲ μηδὲν εἶναι κατὰ δύναμιν, οὕτω δὴ πᾶν ὅσον ἦν ὁρατὸν παραλαβὼν οὐχ ἡσυχίαν ἄγον ἀλλὰ κινούμενον πλημμελῶς και ἀτάκτως, εἰς τάξιν αὐτὸ ἤγαγεν ἐκ τῆς ἀταξίας, ἡγησάμενος ἐκεῖνο τούτου πάντως ἄμεινον.

[31] Indeed, according to Liddell and Scott's Greek lexicon, one sense of *taktos* is "a fixed quantity."

are numerically distinct, but they are qualitatively identical and in that sense are "one."[32] At the genesis of the cosmos, chaotic motion came to be "motion-according-to-number." So even though the "according to number" qualification is an ineliminable and crucial element of Plato's definition of time, he does not go so far as to define time as a kind of number.

We are now prepared to consider some of the difficulties Plato's definition of time faces. At the opening of this section I gestured toward passages where Aristotle makes mention of Plato's cosmology. Examining some of them in light of the preceding discussion will aid us in coming to see how these philosophers' respective conceptions of time might differ from our own. According to Aristotle, Plato commits himself to the following two propositions:

(a) Motion is eternal. (*Met.* 1071b31–3)
(b) Time was created. (*Phys.* 251b14–18)

These are consistent with the textual evidence we have been considering from the *Timaeus*. It is tempting to think that these propositions entail the following claim, which, as we have also seen, Plato explicitly advocates in the *Timaeus*:

(c) Motion existed before time was created.

This claim is not an explicit contradiction, but it certainly seems absurd. For first of all, the very form of the claim invites objection: the *being before* relation obtains between items that are arranged in a temporal series, which presupposes the existence of the item (viz., time) that putatively came into existence only later. So (c) seems to express deep confusion about the logical nature of the relation invoked in the proposition. Second, the item that is claimed to have existed before the creation of time (viz., motion) seems itself to require the existence of time, for how could anything move outside of time? Consequently, (c) appears to be doubly dubious.

And yet Aristotle isn't at all impressed by its apparent absurdity. As it happens, Aristotle rejects the proposition expressed in (c), but not on grounds of absurdity; indeed, he goes to some trouble to show that it is false.[33] We shall have to wait to see what his argument is in Chapter 3, but the mere fact that we find (c) to be absurd, while Aristotle and Plato do not, gives us a compelling (if only prima facie) reason to think that both

[32] I shall show in Chapter 11 how Aristotle inherits and refines this idea of "oneness in form" for movements and times.
[33] Cf. Vlastos (1965), 386.

philosophers are working with a conception of time that is very different from our own. Our discussion in Chapter 1 can help us to identify the salient differences.

It seems to me that the most obvious question that arises from Plato's definition of time was asked immediately above: how could motion (disorderly or otherwise) exist in the absence of time? After all, movements and other sorts of events are items that are extended in such a way that it is practically impossible for us to imagine any particular movement or event without imagining it as being temporally extended. Doesn't this show that Plato must envision the pre-temporal universe as being temporal after all?

Not by his definition. This shows only that he envisions that there was *temporal extension* (or its disorderly analogue) before time was created. Because temporal extension is included as a proper part of our concept of time, this claim strikes us as bizarre; however, the claim is not logically inconsistent, given Plato's definition of terms, and it would be question-begging for us to insist that he is confused on the subject. Though Plato's concept of time is not utterly different from ours, it certainly is different. What we can say at this point is that Plato takes the existence of temporal extension to be a metaphysically necessary condition for the existence of time, though it is not sufficient for the existence of any aspect of time, properly speaking. It is the last point, I gather, with which we would take issue. But for Plato, temporal extension is no *part* of time considered as such, because time is essentially a very specific kind of motion.

This leaves us with two remaining concepts against which to test Plato's definition of time: temporal passage and periodicity. Recall that temporal passage is that element of our concept of time in virtue of which we can speak of the "flow" of the present moment from the past into the future. Periodicity, on the other hand, is that element of our concept of time representing the way in which time presents itself episodically, as being structured so as to have regularly recurring intervals, patterns of passage. Timaeus addresses both of these temporal concepts in the course of presenting his account of the birth of time:

> These [days, nights, months, and years] are all parts [*merê*] of time, and *was* and *will be* are forms [*eidê*] of time that have come into being. (37e3–4)[34]

In light of the preceding account of Plato's definition of time in the *Timaeus*, the claim that the parts of time came into being makes good

[34] ταῦτα δὲ πάντα μέρη χρόνου, καὶ τό τ' ἦν τό τ' ἔσται χρόνου γεγονότα εἴδη.

sense. Time, on Plato's view, is the regular, periodic motion of certain heavenly bodies. In constructing the image of eternity, the demiurge "brought into being the Sun, the Moon and five other stars ... to set limits to and stand guard over the numbers of time" (38c5–6).[35] The very movements of these "wanderers" constitute the periods of natural time. Thus, Plato's concept of time consists at least partly in the concept of periodicity.

But what about temporal passage? Plato's claim that "was" and "will be" came to be alongside the natural periods of time is puzzling. Timaeus' account of the creation of the cosmos is laden with temporal adjectives, and he explicitly commits himself to the temporal ordering of the creative events – in terms of priority and posteriority – at several points (e.g. 34b10 ff.). Since the ordering is expressed in the static terms belonging to the *B* series, a conservative reading of these remarks merely commits Plato to the view that a primitive form of temporal extension existed prior to the birth of time, which is a feature of his view that we've already examined. Still, part of Plato's larger cosmogonic account seems to commit him to temporal passage, and this raises concerns about his claim that "was" and "will be" are coeval with time.

To make the issue even more pointed, let us consider it within the context of the fundamental ontology operating within the *Timaeus*. At the most general level of description, Plato's creation account involves three items:

> Let this, then, be a summary of the account I would offer ... There are being, space, and becoming [*on te kai chôran kai genesin*], three distinct things which existed even before the universe came to be. (52d2–4)[36]

"Being" is meant to designate Plato's forms, the models (*paradeigmata*) after which the structured cosmos is fashioned, and is sometimes characterized metaphorically by Timaeus as "the father." I shall say more about space (the "receptacle" or "wet-nurse") in a moment, but it will be enough for now to note that it is *that in which* the items referred to collectively as "becoming" exist. In the present passage, "becoming" can only refer to the ungenerated raw materials that the demiurge works over in creating

[35] ἥλιος καὶ σελήνη καὶ πέντε ἄλλα ἄστρα, ἐπίκλην ἔχοντα πλανητά, εἰς διορισμὸν καὶ φυλακὴν ἀριθμῶν χρόνου γέγονεν· The *planêta* ("wanderers") are so called both because their trajectories across the night sky are independent of the fixed stars and because some of them (notably Mercury and Venus) exhibit retrogradational motion relative to the other planets – a fact mentioned at *Ti.* 38d2–4.

[36] Οὗτος μὲν οὖν δὴ παρὰ τῆς ἐμῆς ψήφου λογισθεὶς ἐν κεφαλαίῳ δεδόσθω λόγος, ὄν τε καὶ χώραν και γένεσιν εἶναι, τρία τριχῇ, καὶ πρὶν οὐρανὸν γενέσθαι.

the ordered cosmos. Because these raw materials are fashioned into an image of the forms, Timaeus sometimes refers to them as "the child." The metaphor is apt insofar as objects in the perceptible realm are supposed to stand in some sort of resemblance relation with – and to owe their very existence as members of their respective kinds to – those forms of which they are instances. Nonetheless, the metaphor limps to the extent that the metaphysical and epistemological status of the two classes is as different as can be: a form is the kind of thing "which always is and has no becoming … is grasped by understanding … [and] is unchanging," while a particular instance of a form is the kind of thing "which becomes but never is … is grasped by opinion, which involves unreasoning sense perception … [and] comes to be and passes away, but never really is" (27d6–28a4).[37] Children rarely are so different from their fathers.

Metaphors aside, it looks as though Plato intends "becoming" to cover two distinct categories of existents: the unformed raw materials existing in the receptacle prior to the creative acts of the demiurge, and the structured material objects inhabiting space after those acts. There is no reason to object to the equivocal use of the term as such, but Plato's position is precarious. For on the one hand, he claims that "was" and "will be" are forms of time that came into existence when the demiurge created time. Because the ordering relations of *past*, *present*, and *future* belong to temporal passage, Plato's claim suggests that temporal passage is an essential feature of his concept of time. On the other hand, he insists (as evidenced by the passage above) that items subject to becoming existed prior to, and independently of, the creation of the cosmos. By characterizing these items in terms of their coming into and passing out of existence (*gignomenon kai apollumenon*), one might take him to be invoking temporal passage. But surely he can't have it both ways. Ought he, then, to abandon temporal passage as an essential feature of his concept of time, or ought he to give us an account according to which coming into and passing out of existence don't require temporal passage?

Perhaps there is a third way here, one that is roughly coordinate with his views concerning the generation of the elements and time itself. It would not be obviously implausible for Plato to claim that prior to the intervention of the demiurge, there existed a disordered, primitive sort of temporal passage, one that was somehow insufficient to support a

[37] τί τὸ ὂν ἀεί, γένεσιν δὲ οὐκ ἔκον, καὶ τί τὸ γιγνόμενον μὲν ἀεί, ὄν δὲ οὐδέποτε; τὸ μὲν δὴ νοήσει μετὰ λόγου περιληπτόν, ἀεὶ κατὰ ταὐτὰ ὄν, τὸ δ' αὖ δόξῃ μετ' αἰσθήσεως ἀλόγου δοξαστόν, γιγνόμενον καὶ ἀπολλύμενον, ὄντως δὲ οὐδέποτε ὄν.

stringent distinction between past and future. If he were to take this line, he could claim that there were prototypical forms of past and future that "possessed certain traces of what they are now," while still maintaining that they lacked the proper structure to be properly so called.

His argument to this conclusion might run along these lines. Temporal passage seems to consist in the continuous projection of a primitive temporal property – namely, being absolutely present – along the temporal continuum of events. It is the "flow" of the present instant, the now. In order for there to *be* a uniquely present instant, there must be a well-defined simultaneity relation among motion tokens. After all (as Aristotle would agree), the absolute present is a cut across simultaneous events.[38] And, clearly, the existence of the now is a necessary condition for the existence of the future and the past (in whatever sense they can be said to exist): the future just is what lies after the present, and the past is what lies before. Therefore, if there is no well-defined simultaneity relation across individual movements, there is no future and no past – no "was" or "will be."

These considerations suggest that the very distinction between the pre-temporal chaos and post-temporal cosmos can be cast in terms of a well-defined simultaneity relation. In creating time, the demiurge effectively constructs a clockwork.[39] By taking the "discordant and disorderly" movements of the pre-temporal chaos and working them into the numerically regimented revolutions of the celestial spheres, the demiurge quite literally synchronizes the movements, producing a harmonious whole. The result of this act of synchronization is a uniquely present instant, and consequently a genuine distinction between future and past. While it may have been true that each of the disorderly movements lurched along according to its own peculiar "temporal continuum," the lot of them didn't fit together in any rational way. These individual lurchings are what we should regard as the "certain traces" of what was to become genuine temporal passage.

The results of this examination, then, are these. According to Plato's theory, time is fundamentally periodicity, the regular and recurring episodes in the passage of time which are consequent upon celestial motion. The passage of time itself, interestingly, is claimed to depend upon the fitting together of individual movements that constitutes the celestial

[38] *Phys.* IV.13. Aristotle's "now" is a central topic of discussion in Chapter 12.

[39] Cf. Cornford (1937), 102–3. Mohr (1986), 39, also takes this line: "[W]hen Plato says that his craftsman god makes time ... he means that the Demiurge makes a clock, nothing more, nothing less."

clockwork. Strictly speaking, then, temporal passage is also an essential feature of time, on Plato's theory, though he is free (and rightly ought) to maintain that there was a sort of "proto-passage" associated with the disorderly motion of the pre-temporal chaos. Finally, while temporal extension is not a proper part of Plato's concept of time, it is on his view metaphysically necessary for the existence of time.

This brief reconstruction of Plato's account of time puts us in good shape to move on to Aristotle's account. Before I leave the *Timaeus*, though, I should note one final feature of Plato's view. While it is true that the demiurge created all of the finished products that inhabit physical space, it is not the case that he created space itself. The receptacle is one of the three basic constituents of the cosmos (52d2–4). Consequently, there is a profound difference between the metaphysical status of space and time on Plato's view, a difference that is rejected by Einstein and his heirs. For Plato, space is metaphysically basic; and since the clockwork movements of the celestial bodies occur in space, it is a metaphysically (and perhaps logically) necessary condition for the existence of time. If one is at all attracted to the view that space and time form a sort of manifold (the space-time continuum), this deep distinction between the metaphysical status of space and time is suspect. But as we shall see in the next chapter, Aristotle does not challenge this aspect of Plato's theory.

Indeed, Aristotle inherits far more of his views about time from Plato than has been acknowledged to date. Both philosophers regard time as the result of a certain form (namely, *number*) being imposed onto motion, which is an instance of the notion that figures as the conceptual core of hylomorphism. There are important points of disagreement, to be sure, and I shall address these in due course. But given the fact that I shall be arguing for a hylomorphic interpretation of Aristotle's account of time, some general remarks about the role of hylomorphism within his philosophical views are in order as the final piece of stage-setting.

2.3 HYLOMORPHISM

The range of issues to which Aristotle applies hylomorphic analyses cannot be overemphasized. He invokes hylomorphism in connection with logic, language, biology, physics, psychology, metaphysics, ethics, and political theory. The matter–form distinction – and its cousin, the potentiality–actuality distinction – is *horizontally utile*: there are few (if any) matters of philosophical significance whose treatment isn't enriched by thinking in hylomorphic terms.[40]

But within any one of these disparate subjects, Aristotle's use of hylomorphic analysis is also *vertically utile*: a particular hylomorphic compound can itself serve as the matter for a higher-level compound of greater formal complexity. In the following chapter we shall see how Aristotle exploits this fact in his analysis of motion. L. A. Kosman expresses this vertical utility well in discussing the logic of visibility (that which is potentially seen):

> [V]isibility, like so many of the structures of potentiality that Aristotle finds interesting, is subject to an iteration of the potentiality–actuality distinction; in the dark, the visible is only, as we might say, potentially visible. Light effects the actualization of *that* potentiality; it makes what it shines upon actually visible, and thus potentially seen.[41]

The iterative nature of the potency–act and matter–form distinctions is of tremendous philosophical value to Aristotle, for it enables him to employ a single style of explanation and analysis for phenomena of divergent character and complexity. Consider the following simple illustration.

The elements (earth, air, fire, and water) are uncontroversially examples of Aristotelian matter. When various elements are combined in specific ratios (i.e. are imbued with a certain form), higher-level matter is thereby generated – silver, wood, blood, and bone, for example. These are themselves compounds of matter and form, but they most commonly exist as the matter of yet higher-level hylomorphic compounds – in silver bowls, wooden beds, and human animals, among others. The use of kind-terms and count-nouns here is significant and reflects the fact that Aristotle regards things like bowls, beds, and (most especially) living things as the proper answer to the ontological question "What is there?" But that fact doesn't prohibit further iteration of the matter–form distinction, because Aristotle is willing to count the possession of accidental properties by a substance as a variety of form possession.[42] For example, Socrates serves

[40] Anscombe (1953), 85–6, laments this fact: "[Aristotle] tries to use [the term '*hulê*'] by analogy in all sorts of contexts, to extend its application away from where it is so to speak indigenous. I do not know or understand enough to have a general opinion whether the concept, in these extended applications, is so useful an instrument as Aristotle clearly thought it was. Certainly I feel only impatient when he considers calling units the matter of numbers. Nor, for instance, can I make anything of such an idea as 'place-matter'."

[41] Kosman (1992), 347.

[42] Cf. *Met.* 1029b22–5: "There are composites also in the other predications (for instance, quality, quantity, when, where, and motion each has a subject of sorts)" (ἐπεὶ δ' ἔστι καὶ κατὰ τὰς ἄλλας κατηγορίας σύνθετα [ἔστι γάρ τι ὑποκείμενον ἑκάστῳ, οἷον τῷ ποιῷ καὶ τῷ ποσῷ καὶ τῷ ποτὲ καὶ τῷ ποὺ καὶ τῇ κινήσεἰ]).

as the matter, and the property *being seated* serves as the form for "seated Socrates," a peculiar hylomorphic compound appearing in *Metaphysics* Γ.2 (1004b1–4). We are also introduced in *Metaphysics* Z.4 to a pale man who goes by the name "cloak"; the human animal is the matter of cloak, while pallor is cloak's form (1029b22–8).[43] These may strike us as ad hoc or merely conventional entities, but Aristotle treats human beings as the matter of higher-level hylomorphic compounds in rather more ordinary cases, as well. In *Politics* VII.4, for example, he treats a *polis* as a kind of compound entity whose matter is its human populace and whose form is its constitution-type (i.e. democracy, aristocracy, etc.).

So given the general-purpose character of hylomorphism within Aristotle's philosophical method, my insistence on a hylomorphic reading of his theory of time is rather less controversial than it might initially seem to be. And once I have developed a detailed account of what Aristotle takes the *morphê* or *eidos* of time to be (which is the burden of Part III of this book), I hope that any doubts about its viability will be dissolved.

While I have indicated in a very sketchy way how hylomorphic analyses fit into Aristotle's philosophical method, I haven't yet said anything about the inspiration behind hylomorphism, and I ought to do so. For although I have already suggested that Aristotle's use of this style of analysis and explanation is both horizontally and vertically utile, my case for the hylomorphic interpretation of his temporal theory would be further strengthened even at this early stage of development if it were the case that hylomorphism is not only useful, but also *necessary* for the solution of certain philosophical problems. And indeed it is.

At the time when Aristotle was formulating his views, Eleatic monism had been a subject of philosophical discussion for over a century. Parmenides, the intellectual patriarch of the Eleatics, is the first recorded Western philosopher to have denied the existence of motion and change, and his challenge was taken quite seriously by philosophers in Aristotle's day.[44] The Parmenidean challenge is not merely the claim that, as a matter of contingent fact, motion never occurs, but rather is the more radical claim that the very idea of motion – or more generally, of coming to be – is logically incoherent. In one section of Parmenides' poem (he seems to have fancied himself a poet as well as a philosopher), he argues for the

[43] I'll have much more to say about such "accidental compounds" in Chapter 4.
[44] See, e.g., Plato's *Prm.* and *Sph.*

eternal nature of "it" (the One) on grounds of the principle that *ex nihilo, nihil fit* – nothing comes from nothing:

There is still left a single story / of a way: that it is. On this way there are signs / exceedingly many – that being ungenerated it is also imperishable, / whole and of a single kind and unshaken and complete. / Nor was it ever nor will it be, since it is now, all together / one, continuous. For what birth will you seek for it? / How and from where did it grow? I will not permit you to say / or to think "from what is not"; for it is not to be said or thought / that it is not. (28B8)[45]

Parmenides' argument is complex, but the details needn't distract us. The central, though only implicit, point of this compact argument is that one ought not to claim that *being* arises out of *non-being*, for that is a priori impossible.

Aristotle's solution to the Parmenidean challenge is to reject the two-principle analysis of coming-to-be (non-being vs. being) in favor of an analysis involving three principles. In *Physics* 1.7 he claims that every case of coming-to-be includes something that underlies the change (a *hupokeimenon*) and a pair of contraries. So Aristotle rejects the Parmenidean characterization of change as requiring that (for example) the unmusical comes to be musical, because the more accurate description of the acquisition of musical knowledge makes it explicit that an unmusical *man* comes to be a musical man.[46] So Aristotle is quite willing to accept the Parmenidean principle that *ex nihilo, nihil fit*; he simply denies that the proper analysis of coming-to-be yields the result that it must occur *ex nihilo*. For every case of coming-to-be such-and-so requires a *hupokeimenon*, an underlying subject of change.

The analysis is extremely effective in the case of change with respect to accidental properties, but it raises a puzzle in the case of substantial change. What are the three principles (subject and pair of contraries) in the case of an animal's coming-to-be *simpliciter*? This question carries significant philosophical weight, since Aristotle regards coming-to-be

[45] Μόνος δ' ἔτι μῦθος ὁδοῖο / λείπεται ὡς ἔστιν· ταύτῃ δ' ἐπὶ σήματ' ἔασι / πολλὰ μάλ', ὡς ἀγένητον ἐὸν καὶ ἀνώλεθρόν ἐστιν / οὖλον μουνογενές τε καὶ ἀτρεμὲς ἠδ' ἀτέλεστον· / οὐδέ ποτ' ἦν οὐδ' ἔσται, ἐπεὶ νῦν ἔστιν ὁμοῦ πᾶν, / ἕν, συνεχές· τίνα γὰρ γένναν διζήσεαι αὐτοῦ; / πῇ πόθεν αὐξηθέν; οὔτ' ἐκ μὴ ἐόντος ἐάσσω / φάσθαι σ' οὐδὲ νοεῖν· οὐ γὰρ φατὸν οὐδὲ νοητόν / ἔστιν ὅπως οὐκ ἔστι. Trans. McKirahan (rev. Curd) (1994).

[46] Cf. *Phys.* 190a28–31: "[F]or it is true both that the man comes to be musical from being unmusical and that the unmusical one comes to be musical. That is why we also say the same about the compound: we say both that the musical man comes to be musical from being an unmusical man and that the unmusical man comes to be musical" (καὶ γὰρ ἐξ ἀμούσου καὶ ὁ ἄμουσος γίγνεται μουσικός. διὸ καὶ ἐπὶ τοῦ συγκειμένου ὡσαύτως· καὶ γὰρ ἐξ ἀμούσου ἀνθρώπου καὶ ὁ ἄμουσος ἄνθρωπος γίγνεσθαι λέγεται μουσικός).

without qualification (*genesis haplôs*) as forming the conceptual core of coming-to-be in any sense. If we cannot understand the generation of individual substances, then our apparent grasp of qualified coming-to-be will be illusory.[47]

What, then, are the three principles in the case of a horse's coming-to-be? It won't do to say that the contraries are *not existing* and *existing*, for then the question of the *hupokeimenon* would cause embarrassment. If one says that the *hupokeimenon* is a horse, the analysis commits one to the existence of non-existent horses, which is absurd. Parmenides' conception of *genesis* may have been flawed, but he was right to balk at talk about non-existent entities.[48]

Evidently, the better thought is that there is some underlying subject other than the horse that goes from being a non-horse to being a horse. But what is that subject? Aristotle's initial answer is this:

> However, substances – the things that exist without qualification – also come to be from some subject. This will become evident if we examine it. For in every case there is something that is a subject from which the thing that comes to be comes to be, as plants and animals come to be from seed. (*Phys.* 190b1–5)[49]

While this blithe response might answer the immediate question concerning the *hupokeimenon* of a horse's coming-to-be, it raises an identical question concerning the seed's generation. And certainly Aristotle isn't willing to postulate the existence of seed-seeds.

His more considered response is developed in the remainder of *Physics* I.7, where he introduces hylomorphism: "It evidently follows that everything comes to be from the subject (*hupokeimenou*) and the shape (*morphês*)" (190b19–20).[50] The subject in cases of substantial generation is *hulê*, matter:

> The nature that is subject is knowable by analogy. For as bronze is to a statue, or wood is to a bed, or as the shapeless before it acquires a shape is to anything else that has a shape, so the nature that is a subject is to a substance, a this, i.e., a being. (191a7–12)[51]

47 Substantial generation is temporally and metaphysically prior to qualitative change, growth, and decay without exception. It is prior to locomotion in the case of generated beings, but posterior to the eternal movements of the celestial bodies (cf. *Phys.* VIII.7).

48 Cf. *Phys.* I.8, 191b13–29.

49 ὅτι δὲ καὶ αἱ οὐσίαι καὶ ὅσα ἄλλα ἁπλῶς ὄντα ἐξ ὑποκειμένου τινὸς γίγνεται, ἐπισκοποῦντι γένοιτο ἂν φανερόν. ἀεὶ γὰρ ἔστι ὃ ὑπόκειται, ἐξ οὗ τὸ γιγνόμενον, οἷον τὰ φυτὰ καὶ τὰ ζῷα ἐκ σπέρματος.

50 ὅτι γίγνετια πᾶν ἔκ τε τοῦ ὑποκειμένου καὶ τῆς μορφῆς.

51 ἡ δὲ ὑποκειμένη φύσις ἐπιστητὴ κατ᾿ ἀναλογίαν. ὡς γὰρ πρὸς ἀνδριάντα χαλκὸς ἢ πρὸς κλίνην ξύλον ἢ πρὸς τῶν ἄλλων τι τῶν ἐχόντων μορφὴν τὸ ἄμορφον ἔχει πρὶν λαβεῖν τὴν μορφήν, οὕτως αὕτη πρὸς οὐσίαν ἔχει καὶ τὸ τόδε τι καὶ τὸ ὄν.

One simply cannot understand substantial generation independently of the idea of matter taking on shape or form. And since the focal meaning of "*genesis*" is associated with substantial generation, we should expect this fact to carry over into the penumbral cases: as Aristotle says, "it is clear from what has been said that in every case, what comes to be is composite: there is something that comes into being and something that comes to be this" (190b10–12).[52] Thus, hylomorphism is more than a convenient methodological heuristic; it is an indispensable, substantive philosophical doctrine about the nature of anything whose origins are not a matter of strict necessity.[53] As Richard Sorabji succinctly expresses the point, "[Aristotle] is not an anti-realist, who believes that explanations are helpful devices which need not correspond to the real nature of things."[54]

Aristotle, like Plato before him, does not regard the existence of time as a matter of strict necessity. In *Physics* IV.14 he considers whether time would exist if there were no souls, and he endorses a negative response.[55] So time is one of those features of reality whose "origin admits of being otherwise,"[56] and therefore ought to submit to hylomorphic analysis.[57]

I hope that this brief argument serves to motivate the interpretation to be developed in the remainder of the book. As I said in the Introduction, the best argument for the hylomorphic interpretation of Aristotle's temporal theory is not, as the one just given is, couched in generalities. Rather, the best argument in its favor is its ability to make good sense of the text, to cohere with his other doctrines and views, and to weather objections on which other interpretations founder. To see whether it is capable of meeting these demands, we must look at the interpretation in detail.

52 δῆλον ... ὅτι τὸ γιγνόμενον ἅπαν ἀεὶ συνθετόν ἐστι, καὶ ἔστι μέν τι γιγνόμενον, ἔστι δέ τι ὃ τοῦτο γίγνεται.

53 Note, for example, that Aristotle's unmoved movers have neither any material constituent nor any genesis. See chapters 6 and 10 of *Phys.* VIII.

54 Sorabji (1992), 198.

55 This particular claim will be examined in detail in Chapter 3.

56 This phrase appears in *NE* VI.1.

57 To anticipate an objection from the reader familiar with this part of the *Phys.*: Aristotle says in IV.14 that if there were no counting souls, motion could still exist, but time would not. One might object to the line of argument I take here by appealing to Aristotle's (apparent) view that the existence of counting souls is a matter of necessity. (See, e.g., *GA* 731b24–32a1.) Yet even if this is correct, it does not follow that he is committed to the necessary existence of time. For, first of all, the claim in IV.14 seems to be *necessarily, if time exists, then counting souls exist*, and this claim does not in concert with *necessarily, counting souls exist* entail the necessary existence of time. Second, I take it that the necessity of the consequence expressed in IV.14 *itself* captures the relevant notion expressed by "origin admits of being otherwise," since time's existence is alleged to depend upon something other than itself.

PART II

The matter of time: motion

CHAPTER 3

Time is not motion

The hylomorphic interpretation of Aristotle's theory of time quite naturally presents a systematic approach by dividing the task into two stages: first, one should investigate the matter of time, and then one may inquire into the form of time. The burden of this section of the book is to get a conceptual grip on the *hulê* of time, which is *kinêsis*, motion.

Here there are a number of issues and questions to be addressed. The first is to sort out Aristotle's claim that time is an aspect of motion (*kinêseôs ti*). Ursula Coope has produced a reconstruction of Aristotle's argument to this conclusion – which figures as the opening move of *Physics* IV.11 – that deserves careful examination. While her reconstruction posits Aristotle's use of a methodological principle familiar from other treatises,[1] I do not believe that she has got the argument in IV.11 right. What I take to be the correct reconstruction of the argument lights the way for the hylomorphic interpretation. I develop my reconstruction in section 3.2 below.

Independent of the question how to understand this particular argument is the larger concern that Aristotle's definition of time is patently circular. How is it possible to read "a number of motion with respect to the before and after" in such a way as to render it non-circular, let alone informative? Common sense suggests that the definition is bound to fail insofar as the definiens contains terms whose meanings are ineliminably temporal in character: "motion," "before," and "after" are each such that their sense seems to invoke time. This line of thought is by far the most obvious and most common objection to Aristotle's temporal theory.[2]

[1] See Coope (2005), 37–41.

[2] W. D. Ross claims, for example, that Aristotle's remark in *Phys.* IV.14 that were there no counting souls, motion might exist in the absence of time "is clearly unsatisfactory, for obviously change not only could not be apprehended, but could not exist, in the absence of time" (1936, 68). The charge is by no means limited to ancientists, however. Thus John Earman and Richard M. Gale: "[L]ike all attempts to encapsulate the essence of time in some neat formula, [Aristotle's definition is] unhelpfully circular because [it] employ[s] temporal notions" (1995, 803). So too W. H. Newton-Smith: "Aristotle's definition of 'time' as 'the number of motion' is … circular,

The objection is certainly too hasty, and I believe that it is ultimately mistaken. Aristotle defines motion in Book III of the *Physics* as "the actuality of what is potentially, *qua* such,"[3] and there is no overt mention of time here. The principal challenge we face is to clarify the sense of the definition and confirm that there is no surreptitious invocation of temporal notions. I attempt to provide such a clarification and confirmation in Chapter 4.

But even if that challenge is met, the residual problem of "before" and "after" remains.[4] In some ways, this is the fundamental hurdle that must be cleared if one is to appreciate Aristotle's views on time, and for two reasons. First, it is simply not at all obvious how one could understand relations of priority and posteriority, as they relate to motion, independently of any appeal to time. The second reason is that failing to see precisely what Aristotle means by "the before and after in motion" prevents one from seeing that Aristotle is in fact giving a hylomorphic analysis of time. Indeed, I am convinced that the failure of commentators to appreciate the hylomorphic nature of Aristotle's temporal theory is best explained by their failure to understand this crucial phrase (and the underlying notion) within the theory.

On my own interpretation, "the before and after in motion" denotes what I call "kinetic cuts." In intuitive terms, kinetic cuts are instantaneous states of objects in motion. They are, in fact, the matter of "nows," instants of time. But getting a clear understanding of the nature of kinetic cuts is no simple matter. Specifically, it must be established: (a) how they could stand in intrinsic (i.e., non-temporal) relations of priority and posteriority, (b) that Aristotle's commitment to their existence is consistent with his other views on motion, and (c) that they are suitable matter for the requisite form of time. I shall address (a) and (b) in Chapter 5. The issue in (c) will be treated at length in various chapters in Part III.

First, however, it will be instructive to examine very briefly the way in which Aristotle opens his discussion of time, since the puzzles he uses to motivate his own account are revealing and will serve as a kind of test for the hylomorphic interpretation very near the end of the book.

motion being defined itself by reference to change of location in space with regard to time" (1986, 24). The objection is too pervasive to enumerate all of its advocates.

[3] ἡ τοῦ δυνάμει ὄντος ἐντελέχεια, ᾗ τοιοῦτον, κίνησίς ἐστιν (201a10–11).

[4] This objection is articulated clearly by G. E. L. Owen, Julia Annas, and Dennis Corish, whose views I address in section 5.2 below. I say "remains" rather than "remain" because there are not two problems here, but one: to explicate a non-temporal sense of "before" is *ipso facto* to do the same for "after."

Aristotle opens his discussion of time in *Physics* IV.10 by proclaiming that there are two issues to be addressed with respect to the subject: "whether it is among the things that exist or the ones that do not, and then what its nature is" (217b31–2).[5] This is an admirable way to begin, for these have figured even in the twentieth century as central issues in the philosophy of time.[6] Despite the smooth launch, though, the discussion seems to go off course immediately. For instead of considering what he has in mind when he uses the word "time," Aristotle straightaway tackles the question concerning its existence. The sentence following the one just cited reads: "That it either is utterly non-existent or exists scarcely and obscurely might be supposed from the following" (217b32–3).[7] Setting aside the question of what it might be for something to exist "scarcely and obscurely," this way of approaching the pair of questions will seem perfectly backwards to a modern sensibility, since it is usually thought that the reference of a term is determined by its sense. For example, if Smith asks Jones whether he has any *quidsomes* in his pantry, Jones would have no way of knowing how to respond without understanding what Smith means by "quidsome." Unable to attach any sense to the term, Jones is in no position to determine whether it actually refers to anything in his pantry, or anywhere else for that matter. How could Aristotle have made such a serious blunder?

Any felt incredulity should vanish when one sees that the quidsome case is altogether unlike the case of time. Whereas Jones might not have any idea what Smith means by "quidsome," Aristotle surely ought to be permitted (as evidenced in Chapter 1) to assume that his readers are experienced enough in time-talk to have fixed in their understanding a meaning that is, if not completely determinate, sufficiently so as to make them susceptible to the grip of skeptical arguments against its existence. And, indeed, this seems precisely to be Aristotle's assumption, as he proceeds by considering and subsequently rejecting three puzzles that either cast doubt on the existence of time or problematize the status of the now (*to nun*), or instantaneous present. By saying that Aristotle rejects the puzzles, I do not mean to suggest that he provides an explicit solution to each of them; indeed, not only does he not offer solutions, he makes no remark whatever about their quality. But given the way that subsequent chapters develop, one can only suppose that Aristotle must have regarded the

[5] πότερον τῶν ὄντων ἐστὶν ἢ τῶν μὴ ὄντων, εἶτα τίς ἡ φύσις αὐτοῦ.

[6] Certainly, irrealists about time are in the minority, but the influence of McTaggart cannot be dismissed.

[7] ὅτι μὲν οὖν ἢ ὅλως οὐκ ἔστιν ἢ μόλις καὶ ἀμυδρῶς, ἐκ τῶνδέ τις ἂν ὑποπτεύσειεν.

puzzles as bits of sophistic mischief whose tricks are easily identified when one properly understands the nature of time.

Thus, Aristotle makes at least two substantive assumptions about his audience very early on in his account: he assumes that they have no problem engaging in time-talk, and he assumes that they (like him) reject the puzzles out of hand. Of course, rejecting something out of hand is far from having a principled reason for rejecting it, and one indication of the adequacy of the hylomorphic interpretation of Aristotle's temporal theory will be its ability to clearly identify the tricks on which the puzzles hang. This is a task to be postponed to Chapter 13.

Now it is typical of Aristotle to introduce his audience to a new topic by considering some of the problems and views of his predecessors, and after rehearsing the puzzles of IV.10, Aristotle concludes the chapter by reviewing several competing accounts of time. Some of them – the view that time is the sphere of the universe, for instance, which was endorsed by the Pythagoreans[8] – he dismisses as too simple-minded to be taken seriously (218b5–9).[9] However, he does give serious, if only brief, consideration to the view that time is identical with change (*metabolê*). Conceived narrowly as the motion of celestial bodies, this clearly engages Plato's account of time in the *Timaeus*. Aristotle does not trouble himself here to consider the view in any detail,[10] but satisfies himself with two very simple arguments in the form of *modus tollens*. He implicitly assumes some version of the principle of the indiscernability of identicals, according to which motion and time would be indistinguishable from one another, were they in fact identical.[11] He then claims that motion and time are distinguishable in two ways: (i) a movement is only with the thing that is moving, but time is everywhere equally; (ii) every movement is faster or slower, but time is neither of these (218b10–18).[12] He concludes at 218b18

[8] Diels (1912), 45 B, 33: Πυθαγόρας τὸν χρόνον τὴν σφαῖραν τοῦ περιέχοντος εἶναι.

[9] ἡ δὲ τοῦ ὅλου σφαῖρα ἔδοξε μὲν τοῖς εἰποῦσιν εἶναι ὁ χρόνος, ὅτι ἔν τε τῷ χρόνῳ πάντα ἐστὶν καὶ ἐν τῇ τοῦ ὅλου σφαίρᾳ. ἔστιν δ' εὐηθικώτερον τὸ εἰρημένον ἢ ὥστε περὶ αὐτοῦ τὰ ἀδύνατα ἐπισκοπεῖν. It should be noted that Aristotle is able to provide some accommodation even to this view, as simple-minded as it seems. See section 11.4.

[10] Recall that, although Aristotle addresses several features of the *Ti.* cosmological account in the *Cael.* (principally in Bks. III and IV), he does not there evaluate Plato's account of time specifically.

[11] Aristotle's discussion of time in *Phys.* IV.10 and his opening claim of IV.11 are framed in terms of generic change (μεταβολή), but he wastes no time in bringing motion (κίνησις) into the picture and, more specifically yet, locomotion (φορά). I shall frame my discussion almost exclusively in terms of motion (see the opening section of Chapter 4 for a more detailed discussion of this terminological issue).

[12] ἡ μὲν οὖν ἑκάστου μεταβολὴ καὶ κίνησις ἐν αὐτῷ τῷ μεταβάλλοντι μόνον ἐστιν, ἢ οὗ ἂν τύχῃ ὂν αὐτὸ τὸ κινούμενον καὶ μεταβάλλον. ὁ δὲ χρόνος ὁμοίως καὶ πανταχοῦ καὶ παρὰ

that time and change are evidently not identical.[13] While these arguments serve to distinguish Aristotle's theory from Plato's at least superficially, their significance cannot be fully appreciated until we have seen more of Aristotle's view, and in order to do this, we must venture beyond chapter 10 of the *Physics*.

3.1 FLESHING OUT THE OPENING ARGUMENT OF IV.11: TWO SCHOOLS[14]

Aristotle's positive account of time begins in chapter 11 of the *Physics*, which opens with a hotchpotch of arguments whose relation to one another is not immediately clear. His first move is to qualify the two arguments in IV.10 against identifying time with motion by conceding that time does not exist without motion, either (218b21).[15] He then offers what appears to be an argument for the concession and then restates the conclusion at 219a1–2.[16] The sole premise offered in support of the conclusion, it seems, is the observation that we perceive time whenever we perceive motion, but never when we don't perceive motion. Tidied up a bit, this argument runs as follows:

The "Exactly When" argument (218b21–219a1):[17]

1. We perceive time when, but only when, we perceive motion.
2. Therefore, time does not exist without motion.

Aristotle announces that we must take this conclusion as a new starting point and that our task is to determine precisely what time has to do

πᾶσιν. ἔτι δὲ μεταβολὴ μέν ἐστι θάττων καὶ βραδυτέρα, χρόνος δ' οὐκ ἔστιν. τὸ γὰρ βραδὺ καὶ ταχὺ χρόνῳ ὥρισται, ταχὺ μὲν τὸ ἐν ὀλίγῳ πολὺ κινούμενον, βραδὺ δὲ τὸ ἐν πολλῷ ὀλίγον· ὁ δὲ χρόνος οὐχ ὥρισται χρόνῳ οὔτε τῷ ποσός τις εἶναι οὔτε τῷ ποιός. Compare these arguments with the one at *Top.* 120a38 that proceeds by a sort of conceptual analysis in terms of genera and differentiae.

[13] ὅτι μὲν τοίνυν οὐκ ἔστιν κίνησις, φανερόν.

[14] Portions of this section are reprinted from *APEIRON: A Journal for Ancient Philosophy and Science*, vol. 37 (2004), by permission of Academic Printing and Publishing.

[15] Ἀλλὰ μὴν οὐδ' ἄνευ γε μεταβολῆς.

[16] ὅτι μὲν οὖν οὔτε κίνησις οὔτ' ἄνευ κινήσεως ὁ χρόνος ἐστί, φανερόν.

[17] ὅταν γὰρ μηδὲν αὐτοὶ μεταβάλλωμεν τὴν διάνοιαν ἢ λάθωμεν μεταβάλλοντες, οὐ δοκεῖ ἡμῖν γεγονέναι χρόνος … ὥσπερ οὖν εἰ μὴ ἦν ἕτερον τὸ νῦν ἀλλὰ ταὐτὸ καὶ ἕν, οὐκ ἂν ἦν χρόνος, οὕτως καὶ ἐπεὶ λανθάνει ἕτερον ὄν, οὐ δοκεῖ εἶναι τὸ μεταξὺ χρόνος. εἰ δὴ τὸ μὴ οἴεσθαι εἶναι χρόνον τότε συμβαίνει ἡμῖν, ὅταν μὴ ὁρίσωμεν μηδεμίαν μεταβολήν, ἀλλ' ἐν ἑνὶ καὶ ἀδιαιρέτῳ φαίνηται ἡ ψυχὴ μένειν, ὅταν δ' αἰσθώμεθα καὶ ὁρίσωμεν, τότε φαμὲν γεγονέναι χρόνον, φανερὸν ὅτι οὐκ ἔστιν ἄνευ κινήσεως καὶ μεταβολῆς χρόνος. ὅτι μὲν οὖν οὔτε κίνησις οὔτ' ἄνευ κινήσεως ὁ χρόνος ἐστι, φανερόν.

with motion (219a2–3),[18] strongly suggesting that he imagines that this initial stage of the argument is more or less complete as stated.

Why should we regard the "Exactly When" argument as sound? Unfortunately, Aristotle is of little help; rather than attempting to justify the inference, he simply offers anecdotal evidence for the truth of the premise. In support of the claim that perceiving motion is a necessary condition for perceiving time (one half of the premise, construed as a biconditional),[19] he recounts a legend according to which sick individuals would go to Sardinia for treatment followed by a five-day period of sleep (218b23–5).[20] Because they do not notice anything going on around them while they are asleep (and, presumably, because they emerge from their long slumber disoriented, perhaps unaware even that they have been sleeping), it doesn't occur to them that any time has passed without their notice (218b21–7).[21] The sleepers "connect up the former now with the later now and make them one, removing what's between for failing to perceive it" (218b25–7).[22]

Strangely, Aristotle does not offer any justification for the other half of the argument's premise until *after* he has restated the argument's conclusion. This, as we shall see, is just one way in which the argumentative structure of the opening passages of IV.11 is a muddle. Be that as it may, the drowsy imagery of Aristotle's first illustration carries over into the second, which is supposed to convince us that perceiving motion is sufficient for perceiving time:

> [F]or even if it is dark and we are not subject to bodily sensations, but some sort of motion occurs in the soul, it immediately seems that some time has also passed together with it. (219a4–6)[23]

While one might question whether these illustrations succeed in establishing the truth of the premise of the "Exactly When" argument, the premise itself is plausible enough. To be sure, we typically do not actively entertain the proposition that time is passing whenever we attend to some particular movement, but we are ordinarily disposed to say that

[18] ληπτέον δέ, ἐπεὶ ζητοῦμεν τί ἐστιν ὁ χρόνος, ἐντεῦθεν ἀρχομένοις, τί κινήσεώς ἐστιν.

[19] In the interests of simplicity, I ignore certain difficulties by referring to the premise of the "Exactly When" argument as a biconditional and to its parts, so understood, as conditionals.

[20] καθάπερ οὐδὲ τοῖς ἐν Σαρδοῖ μυθολογουμένοις καθεύδειν παρὰ τοῖς ἥρωσιν, ὅταν ἐγερθῶσι. Cf. Ross (1936), 597.

[21] οὐ δοκεῖ … γεγονέναι χρόνος.

[22] συνάπτουσι γὰρ τῷ πρότερον νῦν τὸ ὕστερον νῦν καὶ ἓν ποιοῦσιν, ἐξαιροῦντες διὰ τὴν ἀναισθησίαν τὸ μεταξύ.

[23] καὶ γὰρ ἐὰν ᾖ σκότος καὶ μηδὲν διὰ τοῦ σώματος πάσχωμεν, κίνησις δέ τις ἐν τῇ ψυχῇ ἐνῇ, εὐθὺς ἅμα δοκεῖ τις γεγονέναι καὶ χρόνος.

time passes whenever we witness motion, and no sensible person (Plato notwithstanding) would maintain that motion can take place outside of time.

But even if we are willing to grant the truth of its premise, the inferential character of the argument is questionable, to say the least. Surely there must be some logically adequate, suppressed premise that would be acceptable to Aristotle. But what would it be? Two schools of thought have emerged under the influence of this problem.

The traditional school of thought is one I shall call the "verificationist school" and includes the likes of Sydney Shoemaker, Richard Sorabji, and Edward Hussey.[24] According to the verificationists, Aristotle is committed to the following suppressed premise: any postulation of imperceptible temporal intervals is false (or perhaps even meaningless), because we could never have evidence of their existence.[25] If we construe the premise of the "Exactly When" argument in such a way as to imply the claim that any temporal interval free of motion (such as there might be) would be imperceptible, then the supplied verificationist premise licenses the conclusion that there simply aren't any such motionless intervals. Time does not exist without motion.

Ursula Coope has challenged this interpretation, and while I believe that she is right to do so, I do not agree with her own reading of the argument.[26] It will be instructive to examine Coope's treatment of the text in some detail, insofar as the points on which we differ will form the foundation for the hylomorphic interpretation.

Coope's principal objection to the verificationist interpretation is that it renders half of the premise of the "Exactly When" argument superfluous. The supplied verificationist premise needs just the claim that we perceive time *only when* we perceive motion to justify the conclusion; why, then, would Aristotle bother claiming that we perceive time *whenever* we perceive motion? The claim of sufficiency could more easily be written off as a forgivable slip were it not for the fact that Aristotle repeats the biconditional after restating the conclusion of the "Exactly When" argument (219a3–4).[27] The verificationist therefore seems committed to the view that Aristotle either fails to perceive the superfluous nature of the sufficiency claim, or he simply neglects to point out that it is superfluous. Coope rightly maintains that this is an undesirable commitment and that an

[24] Shoemaker (1969); Hussey (1983); Sorabji (1983).
[25] Cf. Sorabji (1983), 75.
[26] Coope (2001); (2005), 37–41.
[27] ἅμα γὰρ κινήσεως αἰσθανόμεθα καὶ χρόνου.

otherwise equally plausible alternative interpretation that accounted for both halves of the biconditional would be far preferable.[28]

An additional objection to the verificationist reading of this argument emphasizes the peculiarity of attributing the suppressed verificationist premise to Aristotle in the first place. Why should we take Aristotle to have verificationist leanings in connection with time when he seems to have them nowhere else? In the absence of some independent reason to attribute such a view to Aristotle (and the verificationists provide none),[29] the exegetical move looks ad hoc.

The alternative interpretation that Coope recommends seeks to evade both of these objections and is the leading example of the second school of thought, which I shall call the "Owenian school." Coope takes Aristotle's argument at the opening of IV.11 to employ a method he employs elsewhere, namely that of arguing from appearances (*phainomema*). As Owen notes in his famous essay,[30] Aristotle treats "*phainomenon*" as an open-textured term, so that arguing from appearances might involve reasoning from empirical observations, reckoning what the majority of people believe, or considering the historical usage of an important term. Coope has something like the second sense in mind: "On [Aristotle's] view, the fact that we make this assumption [that there is no time without motion] in our ordinary judgements provides a prima facie ground for taking it to be true."[31] The *locus classicus* for Aristotle's endorsement of the principle invoked here is *Nicomachean Ethics* X.2:

> Those who object that that at which all things aim is not necessarily good are really speaking nonsense [*me outhen legousin*]. For we say that that which everyone thinks really is so; and the man who attacks this belief will hardly have anything more credible to maintain instead. (1172b36–3a2)[32]

The principle Aristotle employs in this passage is rather stronger than the one invoked by Coope, as Aristotle seemingly indicates that any philosophical judgment that contradicts common opinion is to be rejected out

[28] Coope (2001), 360–1.

[29] This isn't strictly true, since Hussey (1983), 142, looks to IV.14 for support for his supplemental verificationist premise. Coope (2001), 361, objects that this couldn't be right insofar as the discussion in IV.14 presupposes Aristotle's official definition of time, and the definition presupposes the soundness of the "Exactly When" argument. I concur.

[30] Owen (1975b).

[31] Coope (2001), 362.

[32] Oἱ δ' ἐνιστάμενοι ὡς οὐκ ἀγαθὸν οὗ πάντ' ἐφίεται, μὴ οὐθὲν λέγουσιν. ἃ γὰρ πᾶσι δοκεῖ, ταῦτ' εἶναί φαμεν· ὁ δ' ἀναιρῶν ταύτην τὴν πίστιν οὐ πάνυ πιστότερα ἐρεῖ.

of hand, whereas Coope's interpretation requires only that doing so is prima facie mistaken.[33]

So much the better for Coope's purposes. As she reads *Physics* IV.11, Aristotle's commitment to this principle figures in his larger program in the following way: (a) Aristotle wants "to set up a starting point for his inquiry into time ... by assuming that time is essentially related to [motion]";[34] (b) he attempts to justify this starting point by making plausible the claim that time cannot exist without motion *and also* the claim that motion cannot exist without time;[35] (c) the explicit reason provided for these claims is the observation that we perceive time when, but only when, we perceive motion;[36] (d) the judgments attending these noticings "embody" the assumptions that time cannot exist without motion and vice versa;[37] (e) by employing the principle in question, the alleged fact appealed to in (d) effectively satisfies the desideratum specified in (b), and therefore serves to set up the intended starting point identified in (a).

It is item (b) that helps the Owenian school evade the first objection to the verificationist interpretation. By insisting that Aristotle is arguing tacitly for the converse of the explicitly stated conclusion that there is no time without motion, Coope manages to bring the sufficiency claim of the premise of the "Exactly When" argument into the picture. Her reasons for attributing the additional conclusion to Aristotle is that it "is needed for assuming that time is essentially related to motion" and that parity of reasoning seems to commit him to it.[38]

The Owenian school evades the second flaw in the verificationists' interpretation by employing "a principle that is already familiar from elsewhere in Aristotle's work,"[39] rather than one that looks tailor-made for the problematic passage. To the extent that Coope appeals to a well-established tradition concerning one of Aristotle's dialectical methods, it does seem that she stands on firmer ground on this particular point.

I am, however, not particularly confident that either of the two alleged advantages of the Owenian school's interpretation is sufficient to render

[33] It is worth noting that the *NE* X.2 passage does not obviously endorse the Owenian principle in an unrestricted form. Aristotle might intend "that which everyone thinks" to refer to common opinion only about the good. I am happy to grant the unrestricted form of the principle for the sake of argument.

[34] Coope (2001), 362. I bracket "motion" because Coope frames her discussion in terms of change. Since the distinction is irrelevant in this context (see note 11 above), I shall in the interest of consistency replace "change" with "motion" when quoting Coope's work. In the interest of readability, I shall dispense with the brackets in subsequent instances of such replacement.

[35] Coope (2001), 362–3. [36] Coope (2001), 362–3. [37] Coope (2001), 362–3.

[38] Coope (2001), 362–3. [39] Coope (2001), 365.

it the dialectical victor. Consider the first alleged advantage. Why, one should ask, doesn't Aristotle make explicit the claim that motion cannot exist without time, if he were in fact committed to it? Coope's explanation is this:

> Perhaps it is because he wants to draw the former conclusion [that time is essentially related to motion] rather than the latter [that motion is essentially related to time] that he places so much more emphasis on the claim that there is no time without motion than on the claim that there is no motion without time.[40]

It is surely an understatement to say that he places "more emphasis" on the claim that there is no time without motion, since he nowhere says that there is no motion without time. In fact, he denies the claim that it is impossible for motion to exist without time in IV.14.[41] Had Aristotle admitted even provisionally that there is no motion without time, one would have expected him to make some kind of remark about abandoning the idea when he contradicts it in the final chapter of Book IV. But because he makes no such remark, this element of the Owenian school's interpretation unduly strains the text.

With respect to the second alleged advantage, I do not deny that Aristotle argues from appearances elsewhere in the *Physics* (as he does in his discussion of place in the early chapters of Book IV, for example). Nevertheless, it does not seem to me that Aristotle is arguing from appearances at the opening of IV.11. In order to see why, we shall need to examine Coope's account of how the Owenian principle is employed in this context:

> [Aristotle's] view, I believe, is that our practices in judging *whether or not time has passed* embody certain assumptions about the relationship between time and motion. In making these judgements we take it for granted that there is no time without motion.[42]

[40] Coope (2001), 363 n. 13. Coope's claim that the *being essentially related to* relation is asymmetric is puzzling, since strict identity would seem to be one example of essential relation. S. M. Cohen has suggested to me that Coope uses the phrase as a variant of "is defined in terms of," and I believe that he must be correct.

[41] πότερον δὲ μὴ οὔσης ψυχῆς εἴη ἂν ὁ χρόνος ἢ οὔ, ἀπορήσειεν ἄν τις. ἀδυνάτου γὰρ ὄντος εἶναι τοῦ ἀριθμήσοντος ἀδύνατον καὶ ἀριθμητόν τι εἶναι, ὥστε δῆλον ὅτι οὐδ' ἀριθμός. ἀριθμὸς γὰρ ἢ τὸ ἠριωμημένον ἢ τὸ ἀριθμητόν. εἰ δὲ μηδὲν ἄλλο πέφυκεν ἀριθμεῖν ἢ ψυχὴ καὶ ψυχῆς νοῦς, ἀδύνατον εἶναι χρόνον ψυχῆς μὴ οὔσης, ἀλλ' ἢ τοῦτο ὅ ποτε ὂν ἔστιν ὁ χρόνος, οἷον εἰ ἐνδέχεται κίνησιν εἶναι ἄνευ ψυχῆς (223a21–8). Coope herself makes reference to this passage in connection with her discussion of Hussey's version of the verificationist reading (cf. note 29 above).

[42] Coope (2001), 362. Emphasis added.

Coope goes on to quote two familiar pieces of the text relevant to her interpretation and then reiterates her position:

> Whenever it seems that a certain time has passed, a certain motion seems to have passed together with it (219a7–8).
>
> When we do not mark any motion but the soul seems to remain in one indivisible, it follows that we think there is no time. (218b29–32).
>
> Aristotle thinks that when we make these ordinary judgements *about whether time has passed*, we are assuming that there is no time without motion.[43]

Notice that the judgments behind which the putative embodied belief lies are specified as ones concerning the passage of time: one's judgment that a certain interval of time has passed allegedly embodies the belief that there is no time without motion. In the passages Coope cites, Aristotle is expressing the necessity of perceptions of motion for the perception of time or, equivalently, the sufficiency of perceptions of time for the perception of motion. That is, the embodied belief is of the form *if some time seems to have passed, then some motion has occurred*. What, precisely, is this embodied belief supposed to contribute to the treatment of the phenomenon in question?

Coope's answer is this:

> I have spoken of beliefs that "lie behind" or are "embodied in" the judgements we make. But what is it for a belief to be "embodied in" the judgements we make? Aristotle is not claiming that we all explicitly believe that there is no time without motion or motion without time. His claim is only that these beliefs would explain and justify the particular judgements we make about whether or not time has passed.[44]

Thus, the belief that there is no time without motion is supposed to be capable of explaining and justifying one's judgment that some interval of time has passed. But this is not a plausible view to attribute to Aristotle. The embodied belief is rather better suited to explain and justify the judgment about the occurrence of motion: if an agent judged that *some time seems to have passed*, then a dispositional belief that *if some time seems to have passed, then some motion has occurred* might justify his belief that *some motion has occurred*. This way of thinking about the matter certainly accords much better with the text that Coope cites, since it takes perceptions of time to be sufficient for perceptions of motion. An exactly analogous problem plagues Coope's treatment of the putative embodied belief

[43] Coope (2001), 362. Emphasis added.
[44] Coope (2001), 365 n.16.

that motion requires time, as she maintains that this belief lies behind our judgments concerning the occurrence of motion.[45]

This flaw could be remedied easily enough by systematically interchanging the two types of embodied belief. Unfortunately, however, there are other problems for this interpretation. I think it's implausible to maintain that our judgments about motion and time are either acquired or justified inferentially, and I also think that Aristotle would have recognized the implausibility. Our beliefs about both motion and time typically arise directly from experience and typically are justified (when they are) by the same.[46]

Even setting aside that worry, when it is conjoined with her account of the role played by embodied beliefs, Coope's claim that Aristotle argues from appearances to the conclusion that time requires motion *and* the conclusion that motion requires time saddles Aristotle with absurdity. Recall: the belief that time requires motion is allegedly embodied in our judgments concerning the passage of time, and the belief that motion requires time is allegedly embodied in our judgments concerning the occurrence of motion.[47] Embodied beliefs are ones that "would explain and justify" these judgments. Not by themselves, of course: each must take an antecedent judgment concerning motion or time to serve as an initial premise. That is the point, as I understand it, of saying that the conditionals Aristotle trots out reflect "our ordinary thinking." An individual perceives some motion in his soul; he then determines that some interval of time must have passed on the basis of his implicit belief that motion requires time.

But once this feature of Coope's interpretation is recognized, it becomes quite clear that the explanatory scheme she sketches forms a closed figure. For unless our judgments concerning the passage of time are inexplicable, there is some explanation for them. If the embodied belief is not wholly irrelevant to the actual explanation for these judgments, then if the embodied belief *would* explain the judgments about time, it *does*

[45] Although Coope does not explicitly specify the judgments as such, when one considers the context in which she mentions "these ordinary judgements" (363) in light of the foregoing analysis, it's clear that she must mean judgments about motion.

[46] Cf. Cooper's (1986), 52, complaint about the traditional interpretation of Aristotle's treatment of the practical syllogism in *NE* and *MA*: "[H]aving decided to eat chicken, the agent approaches the table, sees the chicken and takes some – the act of perception not providing a piece of information which is then, as it were, detached and used in a little argument issuing in the act of eating, but instead itself forming a link in a psychological chain leading from decision through perception to action."

[47] For simplicity's sake I here retain Coope's pairings of embodied beliefs with associated judgments.

explain them.[48] That is, to make the embodied belief in any way relevant, it must be counted as an element in the explanatory chain for the explicit judgment. So part of the explanation for someone's believing that time has passed is his implicit belief that motion requires time.[49] But that's not the complete explanation. Another part of the explanation is his having perceived some episode of motion. So if the embodied belief is at all relevant to the judgment in question, then the judgment is formed (in part) on the basis of another judgment concerning the occurrence of motion. As I indicated above, this explanation strikes me as specious; but because Coope also maintains that the explanation for one's belief that motion has transpired is formed (in part) on the basis of another judgment concerning the passage of time, the explanatory trail loops back on itself: we believe that time has lapsed in part because we believe that motion has occurred, and we believe that motion has occurred in part because we believe that time has lapsed. Since circular explanations aren't genuine explanations at all, Coope is mistaken when she claims that the two allegedly embodied beliefs would explain our judgments concerning time and motion.[50]

To sum up the argument of this section: Coope's interpretation of Aristotle as arguing from appearances involves putative explanatory mechanisms. These mechanisms proceed *from* a belief about motion (or time) and a tacit commitment to an ontological principle concerning the relation between motion and time *to* a belief about time (or motion). But because she thinks Aristotle endorses symmetrical ontological principles, the perceptual beliefs in the explanatory scheme stand both as explanans and explanandum to each other. This view is obviously unsatisfactory. To be charitable to Aristotle, then, we may conclude that Coope must be mistaken in one way or another: either Aristotle is not arguing from appearances at the opening of IV.11, or else he is not committed to the claim that motion cannot exist without time. I shall argue that she is mistaken on both counts.

[48] As Coope herself remarks in the footnote excerpted above (2001, 365 n. 16), "[Aristotle] is saying, rather, that this is a belief which *explains* and *justifies* our judgements" (emphasis added). She refers here to Aristotle's account of voluntary actions in *NE* III.5, but if the reference is intended to illuminate her application of the Owenian principle to the argument in *Phys.* IV.11, the counterfactual formulation used earlier is unnecessarily weak.

[49] Here I employ the revised pairings of embodied beliefs and associated judgments, since this is an attempt to characterize the explanatory mechanism charitably.

[50] Similar problems arise in connection with Coope's claim regarding justification, though matters get more complicated here on account of the controversy surrounding the nature of justification itself. I think the foregoing objection framed in terms of explanation casts sufficient doubt upon Coope's interpretation to motivate the search for a better alternative, so I shall set aside the problems associated with justification.

3.2 A NEW OLD SCHOOL

Despite its inadequacies, there is something right about the verificationist interpretation, and the interpretation I shall endorse bears certain affinities with it. I begin by examining some of the text that surrounds the "Exactly When" argument. After stating the initial conclusion that time does not exist without motion and reiterating the point that we perceive motion and time together (219a3–4), Aristotle draws a more specific conclusion than that of the "Exactly When" argument:
The "Something of Motion" argument (219a2–10):[51]

1. And so, time is either identical to motion or is something of motion (*kinêseôs ti*).
2. Time and motion are not identical. (Cf. 218b18.)
3. Therefore, time is something of motion.

Aristotle makes the claim that functions as the first premise at 219a8: *hôste êtoi kinêsis ê tês kinêseôs ti estin ho chronos*. The inferential particle (*hôste* – "and so") suggests that this is a consequence of some earlier claim, but Aristotle has added nothing to the mix prior to making this assertion; he has simply reminded us that time and motion are perceptually concomitant and rehearsed his second illustration in support of the premise of the "Exactly When" argument. So on what grounds does Aristotle make the claim?

I want to suggest that the "Something of Motion" argument is in fact part of the "Exactly When" argument, though Aristotle gives us precious little assistance in recognizing that this is the case. I read the inferential particle as indicating that the first premise of the "Something of Motion" argument is intended to follow from the premise of the "Exactly When" argument. For consider just the form of the "Exactly When" argument:

1. We perceive x when, and only when, we perceive y.
2. Therefore, x does not exist without y.

If the "Something of Motion" argument could be fashioned into the additional premises that are required to render the "Exactly When" argument cogent, two problems would thereby be solved: the otherwise lame

[51] ληπτέον δέ, ἐπεὶ ζητοῦμεν τί ἐστιν ὁ χρόνος, ἐντεῦθεν ἀρχομένοις, τί τῆς κινήσεώς ἐστιν. ἅμα γὰρ κινήσεως αἰσθανόμεθα καὶ χρόνου … ἀλλὰ μὴν καὶ ὅταν γε χρόνος δοκῇ γεγονέναι τις, ἅμα καὶ κίνησιης τις δοκεῖ γεγονέναι. ὥστε ἤτοι κίνησις ἢ τῆς κινήσεώς τί ἐστιν ὁ χρόνος. ἐπεὶ οὖν οὐ κίνησις, ἀνάγκη τῆς κινήσεώς τι εἶναι αὐτόν.

"Exactly When" argument would be made fit, and the apparently spurious "*hôste*" would find a premise to latch on to. In order to determine how the "Something of Motion" argument supplies the required premises for the "Exactly When" argument, we must address the following question: what relations between the values for *x* and *y* in the argument form above could explain the truth of instances of the premise and simultaneously license the corresponding conclusion?

One relation that does the job is identity. If the substituends for *x* and *y* are co-referring terms, then (provided that "perceives" is read *de re*)[52] the instantiated premise of our argument form will be true, and we have before us a very good explanation of its truth. For example, if Phosphorus *just is* Hesperus, then it's obvious both that and why Theophanis perceives (*de re*) Phosphorus exactly when he perceives Hesperus, since there is but one object to be perceived. Additionally, it is also true that Phosphorus does not exist without Hesperus, since each of them is just Venus, and nothing can exist without itself. Hence, if the substituend pair for *x* and *y* are coextensive terms, we have a viable reconstruction of the "Exactly When" argument.

Although identity renders the argument viable, it cannot be the justification that Aristotle has in mind, because he takes himself to have established in IV.10 that time is not identical with motion (see the second premise of the "Something of Motion" argument). Therefore, we must look elsewhere.[53]

The alternative Aristotle considers in the first premise of the "Something of Motion" argument – namely, that time is something belonging to motion – would certainly go a long way towards explaining the fact that we perceive time *only when* we perceive motion, since perceiving the

[52] The *de re*/*de dicto* distinction typically involves the distinction between ordinary objects ("*de re*" means "of the thing") and linguistic or conceptual items ("*de dicto*" means "of what is said"). The distinction between *de re* and *de dicto* perception, as I conceive it, is glimpsed by Aristotle at *Sens.* 446b18–24 and *An. Post.* 100b17. Something is perceived *de re* just in case it originates a causal chain that produces a perceptual event in an agent. The agent might be wholly ignorant of the object's (or property's) character and might even fail to recognize it as the very thing (or kind of thing) that it is. The only requirement is that there is in fact something such that *it* contributes causally, in the typical sort of way, to a perceptual event in the agent. On the other hand, *de dicto* perception implicates the conceptual content of an act of perception under which the agent perceives the intended item. For example, one might perceive the son of Diares variously as a moving white thing, as a man, or as the very man that he is. What one perceives *de dicto* is a moving white thing, a man, and the son of Diares, respectively.

[53] Aristotle's claim regarding the possibility of change without time in *Phys.* IV.14 might be thought to be additional grounds for rejecting identity, but see note 29 above.

features of something plausibly requires perceiving (once again, *de re*) the thing itself.[54]

But surely there are values for *x* and *y*, where *x* is some feature of *y*, and yet it's false that we perceive *x* *whenever* we perceive *y*. For example, if *x* is the molar weight of some chemical compound *y*, there is nothing to guarantee that perceiving the latter strictly requires perceiving the former, in the *de re* sense of "perceiving" or any other sense. Indeed, it would be wildly implausible to maintain that anyone ever perceives the molar weight of a particular chemical compound outside the rather unusual circumstances presented by laboratory experiments.[55] Some features of perceptible objects are such as to be hidden in most contexts.

The very fact that there are such "occult" features invites us to consider whether there might be some that are impossible to overlook. It seems to me that there are in fact such evident features of perceptible objects and that Aristotle probably had just these in mind in his discussion of time, even if he neglected to articulate the idea clearly in the context of that discussion. So I should attempt to be a bit more precise by giving a (provisional) definition of this notion:

(EF) For any pair of perceptibles <*x*, *y*>, *x* is an *evident feature* of *y* iff *x* is a feature of *y*, and for any percipient *z*, *z* cannot perceive (*de dicto*) *y* without also perceiving its *x*.

The idea here is that there are certain features of some items which are such that, should one fail to be aware of those features as such, one wouldn't be aware of the item itself. An example is the pair <volume, stone> (where by "volume" I do not mean any determinate quantity of volume, but simply some volume or other). If Theophanis fails to perceive a particular stone as having volume, it is very difficult indeed to fathom how he could possibly be aware of the stone at all. He could not be aware of the stone *as a stone*, at any rate, since perceiving some *F*-type thing *as* an *F* plausibly requires the employment of the concept of *F*-ness, and any concept of a volumeless object is not the concept of a stone.[56] Of course, the stone isn't

[54] This claim presumes that the feature in question belongs exclusively to the relevant token or type of perceptible item. For example, it's impossible that someone should perceive Callias' pallor without perceiving Callias himself. More on this below.

[55] I assume here that the molar weight of a chemical compound plays no significant role in determining the substance's perceptible features. If this assumption is incorrect (as it might be, for all I know), then by my own account of *de re* perception, one could perceive its molar weight *de re*. Such worries aside, I hope that my point is clear enough. I am grateful to Jim Hankinson for bringing this concern to my attention.

[56] Note that I am not making the implausible claim that in order to perceive a stone as such, one must have an apperceptive belief, part of the content of which is a representation of one's concept

identical with its volume, since it has other essential features (its mass, for example). Another example: <life, animal>. If Theophanis fails to perceive a certain animal as being alive, then he must not perceive the animal *as* an animal, since the concept of an animal (on Aristotle's view, at any rate – cf. *De Anima* 412b18–29) is in part the concept of a living thing.

These two examples illustrate the significance of giving "perceives" a *de dicto* reading, for it's quite possible that an individual should perceive an animal as something other than an animal without perceiving it as something that is alive. (Perhaps he regards it as a very lifelike statue.) My present strategy is to exploit this fact by arguing that Aristotle takes the pair <time, motion> to stand in the evident feature relation, since his doing so would justify the problematic half of the premise of the "Exactly When" argument: if time is an evident feature of motion, then one would in fact perceive time whenever he perceived motion.

Unfortunately, though, the (EF) relation does not justify the claim that we perceive time *only when* we perceive motion. The problem is that the relation places no constraints on the relevant type of feature vis-à-vis its possessor. That is, while any given stone possesses a certain volume, so do elephants and eyeglasses. So while it might well be true that one must perceive volume whenever one perceives a stone as such, one will also perceive volume when one perceives an elephant as such, and it is therefore false that one will perceive volume *only when* one perceives a stone.

The solution to this problem is twofold. Greater care should be exercised in distinguishing types and tokens, and some kind of constraint must be imposed on the relevant types involved. I have been less than cautious in constructing my examples, since the pairs figuring in them are framed in heterogeneous terms. While "volume" and "life" are very commonly used as type terms or mass nouns (one has no hesitation to use either term without any determiner), "stone" and "animal" are both more clearly token terms or count nouns (one speaks indeterminately of "stone" and "animal" in rather special circumstances).[57] I hope that my laxity will be forgiven, however, in light of the fact that Aristotle is rather less careful than he ought to be in his own discussion of motion and time. The premise of the "Exactly When" argument (as well as its conclusion)

of volume. That is, one needn't *think about* one's concept of volume in order to perceive the stone as something possessing volume. All that is required is that one's concept plays a role in one's perceptual belief that a stone lies before one. My appeal to concepts here might well seem unwarranted, but I shall indicate in Chapter 9 how concepts (*empeira*) relate to the form of time.

[57] Note that the grammatical function of "stone" in "stone wall" is adjectival. My claim applies to the nominal use of the terms.

treats "time" and "motion" as type terms,[58] while the illustrations offered as evidence for the premise are framed in terms of token movements and intervals of time. The shift is unobjectionable so long as one makes certain assumptions about uniformity among members of the type and the representativity of the chosen examples. Aristotle apparently makes such assumptions.

When the (EF) relation is augmented to address these concerns, the resulting notion is that of an evident *proper* feature:

(EPF) For any pair of type-perceptibles <Φ, Ψ>, Φ is an *evident proper feature* of Ψ iff every token Ψ_i (and nothing else, except by virtue of its relation to some such token) has some token Φ_i as one of its features, and for any percipient *z*, *z* cannot perceive any Ψ_i *as* a Ψ without also perceiving its Φ_i as a Φ.

The foregoing definition might be thought too contrived to be found anywhere in Aristotle, but in his discussion of properties in *Topics* V.1–3 Aristotle practically illustrates it himself. Consider what he says at 131b33–7:

Thus, for example, a man who has stated that it is a property of a surface to be the primary thing that is colored, has used [in rendering the property] in addition something perceptible, being colored, but something which evidently always belongs, and so the property of surface will in this respect have been correctly rendered.[59]

Given my terminology, this passage indicates that Aristotle regards the pair <color, surface> as falling within the extension of the (EPF) relation. Color and surface are both types each of whose tokens are perceptible (they are type-perceptibles); every surface has some color or other as one of its features;[60] if any non-surface has color, it has it by virtue of being related to a surface (they being the primary colored things); and anyone who perceives a surface as such cannot fail to perceive its

[58] The lack of an indefinite article in Attic Greek complicates the issue, but the absence of "τις" or an alternative determiner clearly indicates that Aristotle intends to be talking about time and motion in general.

[59] οἷον ἐπεὶ ὁ θέμενος ἐπιφανείας ἴδιον ὃ πρῶτον κέχρωσται αἰσθητῷ μέν τινι προσκέχρηται τῷ κεχρῶσθαι, τοιούτῳ δ' ὃ φανερόν ἐστιν ὑπάρχον ἀεί, εἴη ἂν κατὰ τοῦτο καλῶς ἀποδεδομένον τὸ τῆς ἐπιφανείας ἴδιον.

[60] The surfaces of transparent substances (e.g. glass and water) might be thought to present a special challenge; however, Aristotle doesn't hesitate to commit himself to the claim that all bodies whatsoever are colored (*Sens.* 437a6), and in fact his theory of vision maintains that it is the surface of the air at the boundary of our eyes that is the immediate cause of our visual perception of bodies (cf. *An.* 419a11–b1, 423a21–b7, and *Sens.* 439a17–30). More on such topics in Chapter 8.

color – the color is "something which *evidently* always belongs."[61] Other candidates from this section of the *Topics*: <rarity, fire>; <ensouled, living creature>; <anaplestic, liquid>. In each case, an agent who fails to perceive the first member of the pair cannot properly be said to perceive the second member of the pair as such, because *de dicto* perception involves the employment of concepts, and in each case the first member of the pair is a perceptible feature, the concept of which is included in the concept of the second member. A more germane example is the pair <distance, space>, since Aristotle draws a tight connection between space, motion, and time in *Physics* IV.11.[62] This comparison cuts to the heart of Aristotle's temporal theory, and we shall have occasion to revisit it later.

Like identity, the (EPF) relation gives us the means to account for the inference in the "Exactly When" argument. For if time is an evident proper feature of motion, then we will perceive time only when we perceive motion (since it is a *proper* feature of motion), but also whenever we perceive motion as such (since it is an *evident* feature). And so if the "Something of Motion" argument is taken as an argument for the claim that <time, motion> is included in the extension of the (EPF) relation, it

61 The attentive reader ought to have two concerns on this point. First, the scope of "evidently" (*phaneron*) in the passage cited is ambiguous: does Aristotle want to say that it's diachronically evident (i.e. evident upon reflection) that color belongs to surface quite generally, or rather that it's synchronically evident (i.e. evident while viewing a *particular* surface) that color belongs to *it*? While the latter fits more easily into my interpretation, I believe the former alternative is no less serviceable. For if surface and color are both type-perceptibles, and being an evident feature does not require being actively attended to (cf. note 56 above), then its being evident on reflection that color belongs to surface generally would seem quite enough to make it evident at the time of perception *in the relevant sense*. Second, as a variety of the "common perceptibles" (cf. Chapter 9), surfaces are perceptible by means other than sight, so it is strictly incorrect to say that an agent cannot perceive a surface as such without perceiving it as colored. Perhaps Aristotle has in mind his view that sight is the most highly developed sense (cf. *An*. 429a3, *GC* 329b14), under the influence of which one might well say that to perceive a surface in the fullest way possible, one must perceive it as being colored. Further, he holds that we perceive the common sensibles (of which area or size is one) "chiefly" by the faculty of sight (*Sens*. 437a6–9).

62 "Since that which moves moves from something to something, and since every magnitude is continuous, motion follows magnitude; for it's because magnitude is continuous that motion is also continuous, and it's because motion is that time is" (ἐπεὶ δὲ τὸ κινούμενον κινεῖται ἔκ τινος εἴς τι καὶ πᾶν μέγεθος συνεχές, ἀκολουθεῖ τῷ μεγέθει ἡ κίνησις· διὰ γὰρ τὸ τὸ μέγεθος εἶναι συνεχὲς καὶ ἡ κίνησίς ἐστιν συνεχής, διὰ δὲ τὴν κίνησιν ὁ χρόνος) (*Phys*. 219a10–14). Cf. *Met*. 1020a29–32: "[F]or these [motion and time] are called quantities and continuous because the things of which these are attributes are divisible. I mean not that which is moved, but the space through which it is moved; for because that is a quantity motion is a quantity, and because this is a quantity, time is so" (καὶ γὰρ ταῦτα πόσ' ἄττα λέγεται καὶ συνεχῆ τῷ ἐκεῖνα διαρετὰ εἶναι ὧν ἐστί ταῦτα πάθη. λέγω δὲ οὐ τὸ κινούμενον ἀλλ' ὃ ἐκινήθη· τῷ γὰρ ποσὸν εἶναι ἐκεῖνο καὶ ἡ κίνησις ποσή, ὁ δὲ χρόνος τῷ ταύτην). Also cf. *GC* 337a24 ff.

can be incorporated into the "Exactly When" argument, and the result is a rather more plausible argument:

The "No Time without Motion" argument:

1. We perceive time when, and only when, we perceive motion. (218b21–3, 219a3–8)
2. And so (*hôste*), it is either the case that (i) motion and time are identical, or (ii) time is an evident proper feature of motion – since no other explanation for the truth of (1) is available. (Cf. 219a8–9.)
3. But time and motion aren't identical. (218b18)
4. Hence, time is an evident proper feature of motion. (From 1, 2, 3 – cf. 219a9–10.)
5. If *x* is a proper feature of *y*, then *x* does not exist without *y*. (Implicit premise.)
6. Therefore, time does not exist without motion. (From 4, 5 – 218b21.)

I take the plausibility of this argument as a mark in favor of my interpretation. Another mark in its favor as compared to Coope's: this interpretation is much more comfortable with Aristotle's remarks that we perceive motion and time together (*hama* – 219a3) and that we perceive time immediately (*euthus* – 219a6) when we perceive motion, since I claim that they are related objects within a single act of perception, whereas Coope interposes an inference (albeit an unconscious inference) between the perception of one and the judgment that the other must also be present. But though these features of my interpretation are favorable, there are two significant objections that must be addressed before any general claim to its superiority may be made.

The first objection is that my interpretation imports far too much content into the argument, content that should have been made explicit were it intended. I respond, first, by pointing out that without the added content, the argument founders, and as I argued earlier in this section, no other means of repairing the argument on offer is satisfactory. Second, I should like to note that even having imported that content, the argument truly does represent just an opening move, since knowing only that time is an evident proper feature of motion logically entails little more than the fact that time cannot exist without motion. The important question still remains: *in what way* is time an evident proper feature of motion? Aristotle's answer is that time is the number of motion. The explication of this answer occupies Aristotle's efforts for the remaining chapters of Book IV (and will occupy ours for most of the remainder of this book). Compare this with the case of colors and surfaces: one might well be convinced that color is an evident proper feature of surface without having anything like a philosophically respectable account of color.[63] So I insist that, although my interpretation does import a good deal of content into the argument, it does not import too much: the additional content is needed for the argument to go through, is not unfamiliar to Aristotle, and in no way threatens to render the rest of his discussion of time otiose.

The second objection takes the form of a *tu quoque*. I complained in the previous section about the adequacy of Coope's explanation for Aristotle's failure to draw explicitly the conclusion she attributes to him (namely, that there is no motion without time); what explanation can *I* give for the fact that Aristotle claims that there is no time without motion, but *doesn't* claim that there is no motion without time? The brief answer is that Aristotle doesn't believe that time figures in the essence, the "what-it-is-to-be" (*ti ên einai*), of motion – motion is not defined in terms of time.[64] The fuller version of this answer is given in the following chapter, and so I shall have to return to this objection at its end in order to make good on my reply. I am happy to bear the burden of the objection until that time.

To summarize the results of this section: I claim that the interpretation I am advocating is superior both to the Owenian school's and the verificationists', even though my own interpretation entails the latter. On my reading of the opening of IV.11, Aristotle takes time to be a rather specific kind of perceptible feature of motion. According to the verificationists, Aristotle rejects the idea of imperceptible temporal intervals. But if my interpretation is correct, imperceptible temporal intervals would be imperceptible perceptible features, and these are obviously impossible. Thus, my interpretation entails the verificationists'. Still, my interpretation ought to be preferred over theirs insofar as it provides a substantive explanation for Aristotle's rejection of imperceptible temporal intervals and supports both halves of the biconditional in the first premise of the "Exactly When" argument, whereas the verificationists' does neither.

This interpretation is superior to that of the Owenian school insofar as it makes use of the sufficiency claim in the first premise (we perceive time whenever we perceive motion) without attributing to Aristotle the claim that motion cannot exist without time, a claim that he rejects in IV.14 and (as we shall see momentarily) is unsupported by his own account

[63] Note that although Aristotle seems to take color to be an evident perceptible feature of surface in *Top.* V.3, he obviously does not think that this relieves him of the burden of providing an account of color, since he undertakes the provision of just such an account in *Sens.* 3.

[64] Cf. Coope (2005), 9 and 363 n. 12.

of motion. But aside from this fact, my interpretation doesn't saddle Aristotle with an implausible conception of the origins and justification of our beliefs about the passage of time and the occurrence of motion, even while claiming that Aristotle is arguing from appearances.

And so we can now see that Aristotle's first move in laying out his account of time is based on a certain datum of experience: we notice time exactly when we notice motion and change. On the basis of this alleged fact (and other premises drawn from IV. 10 and Book III), Aristotle concludes that time is an aspect of (or "something of") motion – specifically, it is an evident proper feature of motion. But even if one allows that the argument is cogent, the conclusion seems patently absurd on its face. How could time be an aspect of motion, one might ask, when it is quite plain to common sense that motion cannot take place outside of time? Since Aristotle rejects this last notion, it must be that he does not share the common-sense view about motion.

CHAPTER 4

Aristotelian motion

Let us now take up the first charge of circularity, which runs thus: because Aristotle's definition of time includes the term "motion," and because motion cannot be understood independently of temporal notions, the definition cannot possibly succeed in revealing the essential nature of time – rather, it simply presupposes time.

The charge inherits its force from an intuitive conception of motion, according to which motion is just being at different places at different times. Let us follow Wesley Salmon in calling any such view of motion in terms of pairs of spatio-temporal locations a version of the "at-at theory of motion."[1] If Aristotle endorses any form of the at-at theory of motion, it's hard to see how the charge of circularity might be avoided. If he does not, it would be desirable to understand in a fairly precise way what his own view is and how it differs from the at-at family of views.

I should begin by clarifying a terminological issue. I have elected to render "*kinêsis*" as "motion" almost everywhere in this book. But that would be misleading as a strict translation of the Greek term. In the *Categories*, Aristotle draws several distinctions regarding change and motion.[2] *Kinêsis* is one genus of *metabolê*, the highest-level universal of coming to-be. *Kinêsis* includes any variety of change with respect to the states of a subject. The other genus of *metabolê* is change in respect of substance, of which substantial generation (*genesis*) and corruption (*phthora*) are the two species. Each of these two genera was discussed briefly in the concluding section of Chapter 2. The genus in which we are interested has as its three species change in size, with growth (*auxêsis*) and

[1] Salmon (1980), 41–2. Cf. Russell (1903), 469: "The concept of motion is logically subsequent to that of occupying a place at a time, and also to that of change. Motion is the occupation, by one entity, of a continuous series of places at a continuous series of times. Change is the difference, in respect of truth or falsehood, between a proposition concerning an entity and a time *T* and a proposition concerning the same entity and another time *T′* provided that the two propositions differ only by the fact that *T* occurs in one where *T′* occurs in the other."

[2] *Cat.* 15a14 ff.

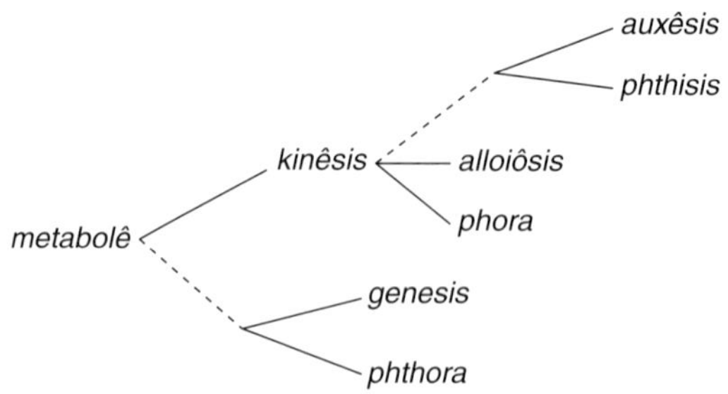

Figure 4. Category tree for *metabolê*.

diminution (*phthisis*) as its two sub-species; qualitative change in respect of color, temperature, and so on (*alloiôsis*); and locomotion (*phora*). This scheme is presented graphically in Figure 4.

Despite its inadequacy as a strict translation, I justify my rendering of "*kinêsis*" as "motion" on three separate grounds. First, Aristotle himself indicates immediately before launching his account of time that it makes no difference whether he speaks in terms of *kinêsis* or *metabolê* (218b19–20), so we are equally safe ignoring substantial change for the time being. Second, in *Physics* VIII.7 Aristotle claims that change with respect to place (i.e. *phora*, locomotion) is the primary form of *kinêsis*, so we may do likewise.[3] Third, it will be far simpler to frame my examples in uniform terms.

Now looking again to the *Categories*, Aristotle remarks there that "what is most distinctive of substance seems to be that something numerically one is capable of receiving contraries" (4a10–11).[4] This capacity of taking on contraries is unique to substances. A color, for example, cannot change in any respect without ceasing to exist (i.e. by being replaced with another color), whereas a man can at one time be pale and at another time not pale, persisting through the change. Colors and other qualities are unfit to serve as subjects of change,[5] although they can serve as the contraries in respect of which a substance (e.g. a particular man) changes.

Aristotle appeals to this way of conceiving of change in the first book of the *Physics*. In I.7 he endorses a tripartite analysis of change according

[3] ὅτι μὲν τοίνυν τῶν κινήσεων ἡ φορὰ πρώτη … (261a27). Cf. *Phys*. IV.1, 208a31–2. The reference is to celestial motion, which will be discussed in Chapter 11.

[4] Μάλιστα δὲ ἴδιον τῆς οὐσίας δοκεῖ εἶναι τὸ ταὐτόν καὶ ἕν ἀριθμῷ ὂν τῶν ἐναντίων εἶναι δεκτικόν.

[5] That is, none is fit to serve as the subject of non-Cambridge change. A particular shade of color can change with respect to popularity, but this is a change in its relational properties, not in its intrinsic features.

to which some underlying subject (*hupokeimenon*) ceases to possess one property and acquires a metaphysically contrary property (191a3 ff.).[6] As we saw at the end of Chapter 2, his account invites certain problems in connection with substantial generation, problems that naturally motivate Aristotle's adoption of hylomorphism. But what is most interesting for our present purpose is the fact that this tripartite account of change seems to have disappeared by the time we reach Book III of the *Physics*, where Aristotle develops his rather more opaque account of change and motion. In the first chapter of Book III, he lays down the following definition: "motion is the actuality of what is potentially, *qua* such" (201a10, restated at 201b4).[7] What has happened to the three "principles" (*archai*) appealed to in 1.7? Does the official definition represent a rejection of the earlier, simpler, account, or is it a refinement?

It is a refinement, though it is not easy to see that this is so. To begin, we would do well to consider a concrete example of motion: say, Diares' sailing from Rhodes to Crete. This particular description of the movement is obviously quite congenial to the tripartite account endorsed in 1.7, since Diares is a paradigmatic candidate subject of change, while *being on Rhodes* and *being on Crete* are suitable contrary properties. How, then, does Aristotle's official definition of motion apply to this toy example?

Insofar as the definition is expressed in terms of actuality (*entelecheia*) and potentiality (*dunamei*), we ought to identify elements in the example to which these notions properly apply. Now before Diares has set sail, he is actually on Rhodes and merely potentially on Crete. Since "actual" and "potential" are ostensibly reciprocal terms – every potentiality is *for* some actuality, every actuality is *of* some potentiality[8] – we should note that Diares' potentiality to be on Crete corresponds with his actually being on Crete. So if "what is potentially" denotes Diares' potentiality to be on Crete, one might be tempted to think that the expression "the actuality

[6] 190b10 ff. See also *Phys.* 188b21–6, *Phys.* 227a7–9. I use the expression "metaphysically contrary" here in order to distinguish what I am discussing from Aristotle's notions of linguistic and logical contrariety. Metaphysical contrariety is a relation that obtains between metaphysical predicates. Bogen suggests that "[*m*]*etaphysical predicates* (e.g. the color red, the relation, to the left of, the action, hitting, etc.) are extra-linguistic items belonging to the classifications Aristotle calls 'categories'. They are signified by *linguistic predicates* (the word 'red', etc.)" (1991, 53). The relation of metaphysical contrariety, then, holds between different metaphysical predicates of the same genus, while linguistic contrariety holds between their linguistic counterparts. Logical contrariety is a relation that holds between types of sentences, e.g. between a universal affirmative statement and its universal negative counterpart.

[7] ἡ τοῦ δυνάμει ὄντος ἐντέλεχεια, ᾗ τοιοῦτον [δυνατόν], κίνησίς ἐστιν.

[8] "Ostensibly" because eternal and imperishable things, which are actualities par excellence, have no potential existence. (Cf. *Met.* 1050b6 ff.)

of what is potentially" in the definition denotes the actuality that corresponds with this potentiality – namely, Diares' actually being on Crete.

This reading is attractive for its simplicity, but it is far too weak. For it implies that while Diares is stationary on Rhodes, *being* on Crete counts as motion for him, which is untenable. Of course, his being on Crete would require that he had moved, but simply being on Crete is not itself motion for Diares, regardless of his present location. For this reason, we cannot read "actuality" and "potentiality" in Aristotle's definition of motion reciprocally – that is, as both corresponding to a single state or property of the subject undergoing motion.

Perhaps the relevant potentiality is Diares' capacity *to be sailing to* Crete, rather than his simply being there. The actuality of this potentiality would then be his actually sailing to Crete, which is of course what the definition is supposed to capture. While this strategy yields the desired result, it has (in the words of Bertrand Russell) all "the benefits of theft over honest toil."[9] Aristotle cannot, on pain of circularity, specify the relevant potentiality in his definition of motion as the potential to be in motion. Another exegetical dead end. What, then, is left to play the role of actuality?

Diares. As a "primary substance" – the individual substances on whose existence, as it is alleged in the *Categories*, the existence of everything else depends[10] – Diares is an actuality without qualification. But it is unclear how we might explicate Aristotle's definition of motion by reference to the subject of change, the *hupokeimenon*. In fact, Aristotle offers an illustration that would seem to block this interpretation:

> What I mean by "*qua*" is this: bronze is potentially a statue. But it is not the actuality of bronze *qua* bronze which is motion. For to be bronze and to be potentially-something are not the same. If they were identical without qualification, i.e. in definition, the actuality of bronze as bronze would be motion. But they are not the same, as has been said. (201a29–33)[11]

This passage highlights the importance of the "*qua*" clause in the definition, and while it explicitly rules out the reading according to which the actuality in question is the *hupokeimenon* considered as such, it invites us

[9] Russell (1903), 71.

[10] "If there were no primary substances, it would not be possible for anything else to exist" (μὴ οὐσῶν οὖν τῶν πρώτων οὐσιῶν ἀδύνατον τῶν ἄλλων τι εἶναι) (2b5–6).

[11] λέγω δὲ τὸ ᾗ ὡδί. ἔστι γὰρ ὁ χαλκὸς δυνάμαι ἀνδριάς, ἀλλ' ὅμως οὐχ ἡ τοῦ χαλκοῦ ἐντελέχεια, ᾗ χαλκός, κίνησίς ἐστιν· οὐ γὰρ τὸ αὐτὸ τὸ χαλκῷ εἶναι καὶ δυνάμει τινί [κινητῷ], ἐπεὶ εἰ τούτὸν ἦν ἁπλῶς καὶ κατὰ τὸν λόγον, ἦν ἂν ἡ τοῦ χαλκοῦ, ᾗ χαλκός, ἐντελέχεια κίνησις· οὐκ ἔστιν δὲ τούτόν, ὡς εἴρηται.

to take seriously the possibility that the *hupokeimenon* figures somehow in the relevant actuality.

4.1 PLASTICITY PROPERTIES AND TELIC PROPERTIES

Recall Aristotle's claim in the *Categories* that "what is most distinctive of substance seems to be that something numerically one is capable of receiving contraries" (4a10–11). As one of Aristotle's primary substances, Socrates is capable of losing his pallor by basking in the sun. Similarly, Diares can relocate from Rhodes to Crete. Socrates' pallor is an accidental – as opposed to essential – property of the man, and the same is true of Diares' being on Rhodes. And yet, as Aristotle tells us in *Physics* II.2, "each has within [him]self a source [*archê*] of motion and stability in place, in growth and decay, and in alteration" (192b13–15).[12] So, whereas Socrates' being pale is an accidental property, his *capacity* to be pale is an essential property, as is his capacity to be dark. One may sensibly ask whether he happens to be dark at a particular time, for that is a matter of historical contingency; to seriously question whether he possesses what it *takes* to be dark is (I dare say) beyond the pale.

So Aristotelian substances possess as part of their very essence an intrinsic capacity, power, or potentiality to take on contrary states.[13] They are, as we might say, essentially *qualitatively plastic*: to be a substance is, among other things, to be the kind of thing that is inherently liable to assuming a great variety of qualitative (and quantitative, and local, etc.) conditions without thereby ceasing to exist or being essentially changed.

The property *being qualitatively plastic* is a second-order property, a property concerning other properties. In being qualitatively plastic, substances have the property of being able to possess contrary properties. Moreover, since it essentially involves the notion of possibility, it is what we would nowadays call a "modal property." But modal properties are traditionally segregated into two different varieties: *de re* ("of the thing"), and *de dicto* ("of what is said"). Aristotle regards a property of the type in question as an intrinsic feature of the object itself (Socrates is himself possibly dark), not a linguistic property belonging to some sentence or proposition ("Socrates is dark" is possibly true). Thus, such properties are

[12] τούτων μὲν γὰρ ἕκαστον ἐν ἑαυτῷ ἀρχὴν ἔχει κινήσεως καὶ στάσεως, τὰ μὲν κατὰ τόπον, τὰ δὲ κατ' αὔξησιν καὶ φθίσιν, τὰ δὲ κατ' ἀλλοίωσιν.

[13] "Capacity," "power," and "potentiality" are equally common translations of δύναμις.

de re modal properties.[14] Let us call all such second-order *de re* modal properties "plasticity properties."

I take it that Aristotle's commitment to the existence of plasticity properties is universally acknowledged, though it might be expressed by others in more prosaic terms. What is perhaps less well understood is the nature of Aristotle's commitment to a different kind of modal property of substances that figures centrally in his account of motion. The failure of commentators to clearly identify these properties and articulate their character is, I believe, responsible for the general confusion surrounding Aristotle's definition of motion. The properties in question are intimately related to plasticity properties but differ in that they are first-order modal properties, rather than second-order, and they are merely accidental, rather than essential.

The variety of first-order modal properties to which I believe Aristotle is committed are ones that I should like to call "telic properties." An example of a telic property is *being potentially located in the agora*. A telic property, like a magnetic charge property, endows its possessor with a kind of intrinsic orientation or directedness within its environment. Magnetic properties are accidental first-order properties, but they also have an essential second-order modal counterpart: by its very nature, an unmagnetized iron needle has the second-order modal property *being possibly magnetized* (or, perhaps, *being magnetizable*). Magnetism is itself not obviously a modal property, but it is an apt comparison for telic properties, and Aristotle seems to conceive of the latter as a variety of modal property (as evidenced by the appearance of "potentially" in their descriptions). I don't think terribly much hangs on the question of whether telic properties are genuinely modal properties or not; so long as we understand the nature of these properties and their second-order modal cousins, we shall find ourselves well equipped to understand Aristotle's definition of motion.

Because the distinction between plasticity properties and telic properties is blurred by Aristotle's language of potentiality, I shall for the purpose of articulating the substantive difference between the two employ instances of "possibly *F*" to designate plasticity properties and reserve instances of "potentially *F*" to designate telic properties. The stipulated linguistic contrast has no exact counterpart in Aristotle's terminology,

[14] For independent evidence that Aristotle has no qualms with *de re* modality, see *Int.* 12 21b11 ff. and *SE* 4 166a 24 ff.

though I believe the theoretical apparatus required by his account makes the distinction inevitable.[15]

As a way of beginning that task, let me pre-empt one line of objection that was all too familiar among critics of scholasticism, according to which Aristotelian definitions are vacuous. Against such objections, I must insist: to attribute a telic property to an object is *not* simply a fancy way of saying that the object happens to be moving, any more than attributing magnetism to a needle is simply a fancy way of saying that it happens to align itself with the earth's magnetic poles. The attribution of magnetism is intended to *explain* the needle's behavior by revealing something about the structure of reality, and similar considerations apply in the case of attributing telic properties. In raising this defense of Aristotle I am not, of course, claiming that telic properties in fact exist. (I happen to have philosophical reservations about such properties.) Rather, I am claiming that attributing telic properties is not the sort of empty word-play portrayed by Molière in "explaining" opium's power to induce sleep by attributing a "dormative virtue" to it.

I shall develop a detailed account of the role of telic properties in the existence of motion below, but as a first step to seeing how attributing telic properties can be philosophically enlightening, let us consider the case of an ordinary (non-modal) first-order property – say, the property *being green*. An object that is actually green cannot also actually be not-green. Possessing one property precludes (on grounds of metaphysical impossibility) the possession of the contrary.[16]

In contrast, being green does not preclude possessing the plasticity property *being possibly not-green*.[17] Indeed, on the plausible assumption

[15] In particular, Aristotle's use of iterated "potentiality" and "actuality" modifiers does not correspond to the distinction between plasticity and telic properties. His use of such iteration in his definition of motion *invokes* telic properties, but the distinction is inexpressible in terms of such iteration. This fact is probably responsible, to a large extent, for the many and varied failed attempts to make good sense of Aristotle's definition.

[16] Note that Aristotle formulates the law of non-contradiction ("the most certain principle") in *de re* terms, rather than *de dicto*: "the same attribute cannot at the same time belong and not belong to the same subject in the same respect" (τὸ γὰρ αὐτὸ ἅμα ὑπάρχειν τε καὶ μὴ ὑπάρχειν ἀδύνατον τῷ αὐτῷ καὶ κατὰ τὸ αὐτό) (*Met.* 1005b19–20). A *de dicto* formulation would couch the law in terms of truth-values and statements, rather than in terms of objects and properties. See note 52 in Chapter 3.

[17] Cf. *Cael.* I.12: "A man has, it is true, the capacity at once of sitting and of standing, because when he possesses the one he also possesses the other; but it does not follow that he can at the same time sit and stand, but rather only at different times" (τοῦ μὲν οὖν καθῆσθαι καὶ ἑστάναι ἅμα ἔχει τὴν δύναμιν, ὅτι ὅτε ἔχει ἐκείνην, καὶ τὴν ἑτέραν· ἀλλ' οὐχ ὥστε ἅμα καθῆσθαι καὶ ἑστάναι, ἀλλ' ἐν ἄλλῳ χρόνῳ) (281b15–18). Also cf. *Int.* 21b33–7: "The negation of 'possible not to be' is 'not possible not to be'. This is why 'possible to be' and 'possible not to be' may be thought actually to follow from one another" (Τοῦ δὲ δυνατὸν μὴ εἶναι ἀπόφασις οὐ

that nothing is essentially green, being green *entails* possessing the plasticity property of being possibly not-green, and essentially so.[18] For being green requires being capable of possessing color properties in the first place; and if nothing is essentially green, then any object that is in fact green will be capable of possessing colors other than green. So, necessarily, any green object possesses a wide array of color plasticity properties: *being possibly red*, *being possibly blue*, *being possibly yellow*, and so on. Second-order modal correlates are co-instantiable, and each is co-instantiable with first-order color properties associated with those correlates.

Likewise, neither does being green preclude the possession of the telic property *being potentially red*. An object that has this telic property is one that stands in an "attractive" metaphysical relation with the property *red*, in much the same way as a magnetized needle stands in an attractive relation with the magnetic poles of the earth, or as two massive bodies stand with one another in gravitational attraction. The property in question, as Aristotle envisages it, is the feature of reality by virtue of which the green object comes to be red – not as the *agent* of change, but rather as being partly *constitutive* of the change as such. The way in which a telic property is partly constitutive of change and motion will be addressed in detail below.

While correlate plasticity properties are co-instantiable with one another, correlate telic properties are not. Nothing can co-instantiate the properties *being potentially green* and *being potentially red*. Our evidentiary basis for affirming this, Aristotle would happily admit, is the fact that we never observe anything "greening" and "reddening" simultaneously. But he would insist that there is a deep metaphysical truth here, one that explains the evidence: the correlate telic properties are such that the instantiation of one precludes the instantiation of the other.[19] In this respect, telic properties are as fully robust as their non-modal first-order counterparts *being green* and *being red*.

Note, too, that while something that is red *ipso facto* has the plasticity property *being possibly red*, it is impossible for a red object to possess the telic property *being potentially red*. The very nature of a telic property is such as to endow its possessor with an intrinsic directedness with respect

τὸ οὐ δυνατὸν εἶναι ... διὸ καὶ ἀκολουθεῖν ἂν δόξειαν ἀλλήλαις αἱ τοῦ δυνατὸν εἶναι καὶ δυνατὸν μὴ εἶναι).

[18] The "plausible assumption" here ignores the sort of compound entities discussed in the following section.

[19] "[I]t is impossible for opposite movements or changes to be present together" (οὐκ ἐνδέχεται ἅμα τὰς ἀντικειμένας οὔτε κινήσεις οὔτε μεταβολὰς ὑπάρχειν) (261b20–2).

to another determinate property. If the object in question actually possesses the determinate property, there is no directedness to be conferred.[20] In this respect, telic properties are (in an accidental way) oriented toward their own destruction.

As helpful as the comparisons with gravitational attraction and magnetism are in getting a conceptual purchase on the nature of telic properties, there are clearly important differences. Gravitational attraction is indiscriminate (*any* two bodies are mutually attractive) and symmetric (any two bodies are *mutually* attractive), whereas a telic property is highly fine-grained (Socrates has the property *being potentially at Plato's door*) and asymmetric (Plato's door has no telic property directed toward Socrates). Magnetism, like gravitational attraction, is also symmetric and additionally happens to be indefinitely sustainable (a needle's possessing magnetic charge can be highly stable), whereas a telic property is self-exhausting.

4.2 ACCIDENTAL COMPOUNDS

Primary substances are hylomorphic compounds, and they are the fundamental entities in Aristotle's ontology. But as I hinted in section 2.3, Aristotle is quite willing to consider the status of composite entities other than primary substances. So, for example, he maintains in *Metaphysics* Γ.2 that it is within the province of "first philosophy" to consider the question "whether Socrates is the same as seated Socrates" (1004b2–3).[21] Here, "seated Socrates" refers to a composite object whose matter is Socrates and whose form is the property *being seated*. If the complex phrase were being used simply as a roundabout way of referring to Socrates independently of his seated condition, the question posed would be perfectly trivial and unworthy of philosophical examination. Insofar as being seated is an accidental property, let us say that "seated Socrates" names a particular *accidental compound*, if it names anything at all.[22]

Aristotle's interest in accidental compounds is fully evident in *Metaphysics* Z. 4, where he not only countenances one he calls "the pale man," but also introduces a singular term to function as a proper name

[20] Compare: a reliable compass does not orient itself in any particular direction when it is located precisely at magnetic north.

[21] εἰ ταὐτὸ Σωκράτης καὶ Σωκράτης καθήμενος.

[22] Recall *Met.* 1029b22–5: "There are composites also in the other predications (for instance, quality, quantity, when, where, and motion each has a subject of sorts)" (ἐπεὶ δ' ἔστι καὶ κατὰ τὰς ἄλλας κατηγορίας σύνθετα [ἔστι γάρ τι ὑποκείμενον ἑκάστῳ, οἷον τῷ ποιῷ καὶ τῷ ποσῷ καὶ τῷ ποτὲ καὶ τῷ ποὺ καὶ τῇ κινήσει]).

for this compound entity: "Let us, then, call this composite 'cloak'" (1029b27–8).[23] Aristotle employs the logically proper name in order to prevent the reader from separating the pallor and the man in thought, as one might be inclined to do when the compound is designated by means of two words. An apparent consequence of Aristotle's willingness to take on board such compounds is the fact that he is prepared to countenance objects that are essentially pale (as surely cloak must be).[24]

Now, such compound entities might strike some as metaphysically bizarre, but Aristotle takes a more liberal stance, viewing them as simply failing to be the *primary* kind of existents. As he says in *Metaphysics* Z.4, "there can be a formula or definition of pale man, but not in the sense in which there is a definition either of pale or of a substance" (1030b12–13).[25] The explanation for this fact is that, being composites of items from disparate ontological categories, there is no single category to which they belong: they are, to borrow a phrase from S. M. Cohen, "kooky objects."[26] Yet despite their mixed character, they are employed by Aristotle in the treatment of substantive philosophical questions. And not only within "first philosophy," but also within "second philosophy," i.e. the study of nature.

Indeed, perhaps the earliest discussion of accidental compounds is in *Physics* 1.7, where Aristotle first articulates his tripartite analysis of motion and change. The proper description of a case of acquiring musical skill and knowledge is this: a man, who formerly was unmusical, comes to be musical. The subject of change (*hupokeimenon*) in this case is the man, and *unmusical* and *musical* are the contrary properties.

Although this is Aristotle's preferred description on the tripartite analysis, another common description says that the unmusical man has become a musical man. The only interpretation on which such a description is not self-contradictory is one where the complex linguistic term "unmusical man" is being used to refer obliquely to the man independently of his being unmusical, and not to the accidental compound per se. The oblique

[23] ἔστω δὴ ὄνομα αὐτῷ ἱμάτιον. Cf. Aristotle's use of the same singular term in *Met.* H.6 to refer to "the round bronze" (στρογγύλος χαλκός). Also cf. *PA* 649b23–4: "blood … is spoken of as boiling water would be were it denoted by a single term" (τὸ αἷμα … καθάπερ γὰρ εἰ ὀνόματι τινι σημαίνοιμεν τὸ ζέον ὕδωρ, οὕτω λέγεται·). Blood is a compound stuff of more basic material constituents in a determinate ratio, but it is common enough to deserve a unique kind-term as its name.

[24] See note 18 above.

[25] διὸ καὶ λευκοῦ ἀνθρώπου ἔσται λόγος καὶ ὁρισμός, ἄλλον δὲ τρόπον καὶ τοῦ λευκοῦ καὶ οὐσίας.

[26] Cohen (2008).

use of a complex referring device like this constitutes a mismatch between the structure of language and the structure of thought, but the mismatch is not objectionable as such. The risk inherent in these cases of mismatch is that one stands to induce confusion in others.[27] In saying that the unmusical man has become a musical man, for example, one risks the appearance of being committed to the view that, at the end of the process, there is a man who is both unmusical and musical. Aristotle's haste in pointing out the illegitimacy of the inference (190a10–13) demonstrates his preference for the strict use of complex referring terms, whereby "the unmusical man" refers to an accidental compound composed of a man and the property *unmusical*. Thus, under ordinary circumstances, "the unmusical man" simply fails to refer when the student has acquired musical knowledge and skill. Analogously, it will be a conceptual or necessary truth that the musical man is musical (when the complex term is used strictly to refer to the accidental compound), whereas it will be a contingent truth that the man who happens to be musical is in fact musical (when "the musical man" used to refer obliquely to the student).

4.3 TELIC PROPERTY COMPOUNDS

Our foray into telic properties and accidental compounds was motivated by a question and an insight. The question was this: to what does "actuality" refer in Aristotle's definition of motion as "the actuality of what is potentially, *qua* such"? The insight was this: probably, "actuality" is in some way related to the subject of change – the *hupokeimenon* – though the passage discussing the bronze statue shows that it cannot be identical to the underlying subject of change.

I suspect that my interpretative strategy is now evident. I claim that Aristotle defines motion in terms of telic property compounds, accidental compounds one of whose constituents is a telic property. This reading is strongly suggested in *Physics* IV.11, where Aristotle says:

> a movement is made one by the thing which is moved [*tô pheromenô*], because it is one (not because it is one in substrate [*ho pote on*] – for that might leave gaps – but because it is one in definition). (220a7–8)[28]

[27] Cf. *Top.* VI.11, 149b33 ff., where Aristotle recommends looking for one-to-one correspondence between parts of a definition and parts of the thing defined. Interestingly, he uses the term "cloak" here for a compound entity (viz., the doublet).

[28] ἡ φορὰ μία τῷ φερομένῳ, ὅτι ἕν (καὶ οὐχ ὅ ποτε ὄν – καὶ γὰρ ἂν διαλίποι – ἀλλὰ τῷ λόγῳ).

Here the English phrase "the thing which is moved" translates a single Greek term, "*pheromenon*," which is a species-term variant of "*kinoumenon*."[29] That expression is intended to refer to a telic property compound – for example, Coriscus compounded with the telic property *being potentially in the Lyceum*. Aristotle goes out of his way to distinguish the subject of change from the *kinoumenon*, maintaining that the existence and unity of the subject of change is insufficient to secure the existence and unity of the movement. The *kinoumenon* and the *hupokeimenon* are "one in number" in the sense that they are co-located; yet they are not the same in definition, for being Coriscus is different from being a telic property compound.[30] (This is precisely Aristotle's point in saying that "to be bronze and to be a certain potentiality are not the same" [201a31–2].)[31] The *kinoumenon just is* what (i.e. *is* the thing that) is potentially in the Lyceum. Hence, cognate instances of the phrase "what is potentially *F*" are equivalent to corresponding instances of "the (*F*-type) *kinoumenon*" in much the same way that "the pale man" is equivalent to "cloak" in *Metaphysics* Z.4.

These insights shed considerable light on Aristotle's definition of motion as "the actuality of what is potentially, *qua* such." The unit-phrase "what is potentially" functions as a generic kind-term for *kinoumena*, and the "*qua* such" qualification makes it clear that the phrase refers strictly to the telic property compound as such, not simply the *hupokeimenon* of the compound. By defining motion as the actuality of a *kinoumenon*, Aristotle means to indicate that motion, as an actual concrete phenomenon, consists in the existence of some telic property compound: for motion to occur is simply for some *kinoumenon* to exist.

One bit of textual evidence in favor of this interpretation is a passage that follows very closely on the heels of Aristotle's definition:

> For each thing is capable of being at one time actual, at another not. Take for instance the buildable: the actuality of the buildable [*oikodomêton*] *qua* buildable is the process of building [*oikodomêsis*]. For the actuality must be either this or the house. But when there is a house, the buildable no longer exists. On the other hand, the buildable is that which is being built. Necessarily, then, the actuality is the process of building. But building is a kind of motion, and the same account will apply to the other kinds also. (201b7–15)[32]

[29] Recall that φορά is one species of κίνησις (see Figure 4, this chapter). I shall opt to use the more inclusive term "*kinoumenon*" hereafter, except where the distinction is significant.

[30] Cf. *Phys.* I.7, 190a17, *Met.* Z.4, 1029b13–16.

[31] οὐ γὰρ τὸ αὐτὸ τὸ χαλκῷ εἶναι καὶ δυνάμει τινί.

[32] ἐνδέχεται γὰρ ἕκαστον ὁτὲ μὲν ἐνεργεῖν ὁτὲ δὲ μή, οἷον τὸ οἰκοδομητόν, καὶ ἡ τοῦ οἰκοδομητοῦ ἐνέργεια, ᾗ οἰκοδομητόν, οἰκοδόμησίς ἐστιν (ἢ γὰρ οἰκοδόμησις ἡ ἐνέργεια

Here, "the buildable" translates *oikodomêton*, which could also be rendered "that which is potentially a house." Understood as a telic property compound (the telic property *being potentially arranged house-wise* instantiated in an assortment of building materials), the buildable as such is a *kinoumenon*. The actuality of this kind of *kinoumenon* just is the process of building. When the *telos* is realized, the telic property is no more, and so the telic property compound – the *kinoumenon* – likewise is no more. This is why Aristotle says that "when there is a house, the buildable no longer exists."

I note a strength of this interpretation as compared to the interpretation offered by Ursula Coope, according to which the "*qua* such" qualification within the definition functions simply to limit the potentialities in question to incomplete ones, so that "[a lump of] bronze's potential to be a statue is actual-*qua*-potential when it is not yet a statue."[33] This approach seems to me not to take seriously Aristotle's attempt to provide a real definition of motion. An interpretation that entails Coope's criterion and also explains why this is in fact a feature (though not a criterion) of motion would certainly be preferable. Coope herself seems to regard her criterion as being unsatisfactory when she laments that "This notion of a potential's being actual is almost impossible to explain without resorting to metaphor."[34] Like so many other commentators, Coope takes "actuality" to modify "potential" within Aristotle's definition of motion.[35] But as we have seen, taking "actuality" to modify the unit-phrase "what is potentially (such)" is a much more natural interpretative choice in that it doesn't reify potentialities, but instead properly places substance at the center of the definition. What's more, we have also seen how the interpretation on offer provides a substantive explanation for Coope's criterion couched in terms of the logic of telic properties (necessarily, no F-type *kinoumenon* is actually F), and it does so without resorting to metaphor.

Now, despite these apparent advantages, the interpretation on offer might seem implausible on its face, for motion is evidently a kind of

ᾗ ἡ οἰκία· ἀλλ' ὅταν οἰκία ᾖ, οὐκέτ' οἰκοδομητὸν ἔστιν. οἰκοδομεῖται δὲ τὸ οἰκοδομητόν· ἀνάγκη οὖν οἰκοδόμησιν τὴν ἐνέργειαν εἶναὶ)· ἡ δ' οἰκοδόμησις κίνησίς τις. ἀλλὰ μὴν ὁ αὐτὸς ἐφαρμόσει λόγος καὶ ἐπὶ τῶν ἄλλων κινήσεων.

33 Coope (2005), 9. Coope appeals to a passage at *Phys.* 201b31 ff. discussing "incomplete actualities" (ἀτελὲς τὸ δυνατόν) in support of this characterization, but (a) the passage is dialectical, and it's unclear whether Aristotle embraces the notion in question (which he attributes to the Pythagoreans), and (b) even if he does embrace it, there's little reason to suppose that he intends it to illuminate his own definition.

34 Coope (2005), 7.

35 I have been equally guilty of this error in the past. See, e.g., Roark (2004), n. 5.

process, not a kind of object. And indeed, the language Aristotle uses in framing his examples pushes us in this direction. Consider the fact that the actuality in the passage above is identified as *oikodomêsis*, which is universally translated as "the process of building." The same term appears in *Nicomachean Ethics* VII.11, there also in connection with process-oriented language: "no case of coming-to-be (*genesis*) is of the same kind as its end, e.g. no process of building (*oikodomêsis*) is of the same kind as a house" (1152b14–15).[36] How can I square this process language with my view that Aristotle defines motion as the actuality of a *kinoumenon*? And perhaps more fundamentally, how can I justify the introduction of telic property compounds, entities that might strike some scholars as rather mysterious?

To the first question, it is worth reiterating Aristotle's motivation for conceiving of motion in the way I have argued that he does. His objective in so defining motion is to remain true to his stated position in the *Categories*, explicitly reaffirmed in *Metaphysics* Γ.2, and sustained everywhere that *being* must always be rooted in particular substances. Thus, not only is it the case that individual movements are always movements *of* particular substances, it is also the case that *motion itself is something of substance*.[37] A particular movement is made one by the existence of particular *kinoumenon* (recall 220a6–8). Therefore, motion considered quite generally is what it is by virtue of the existence of *kinoumena*, where "*kinoumenon*" functions as the kind-term analogue of the singular term "cloak" that Aristotle coins in *Metaphysics* Z.4 to refer to the accidental compound "the pale man." So seeking an interpretation of his definition that privileges substance is not at all suspect.

To the second question, I maintain that the appearance of mystery is neither unexpected nor exegetically undesirable. Admittedly, the existence of motion is patently obvious to any normal human being, and yet telic properties and composites formed from them seem to be rather exotic creatures in the philosophical sanctuary. But to reject the account simply for this reason is to fail to appreciate the character of scientific inquiry as Aristotle conceives it. Consider the remarks with which he opens the first chapter of the *Physics*:

The natural way of [discovering first principles] is to start from the things which are more knowable and clear to us and proceed towards those which are clearer

[36] οὐδεμία δὲ γένεσις συγγενὴς τοῖς τέλεσιν, οἷον οὐδεμία οἰκοδόμησις οἰκίᾳ.

[37] Recall Aristotle's argument in *Phys.* IV.10 where he says that a movement is only with the thing that is moving (218b10).

and more knowable by nature; for the same things are not knowable relatively to us and knowable without qualification. So we must follow this method and advance from what is more obscure by nature, but clearer to us, towards what is more clear and more knowable by nature. (184a15–21)[38]

Given this conception of scientific inquiry, it's not at all surprising that Aristotle's definition of motion would seem to obscure something that antecedently seemed quite clear. For while motion is more knowable to us as an observable phenomenon, it is more knowable *simpliciter* as what it really is: in Aristotle's words, the actuality of what is potentially, *qua* such; in my terminology, the existence of a compound entity whose constituents are an ordinary substance and a certain telic property – or more compactly, the actuality of a *kinoumenon*. The definition takes us from the undisputed phenomena, which are intuitively transparent but theoretically opaque, to the underlying reality, which is intuitively opaque but theoretically transparent.

Evidence for the theoretical transparency of the definition as I've interpreted it is the fact that it has sufficiently clear and structured content to figure in further theoretical development. The specifics of this evidence will be displayed in the remainder of this chapter and in Chapter 5. Before I endeavor to lay out the evidence, though, let's pause to consider how the foregoing interpretation bears on the objection raised at the beginning of this section, which runs thus: given the fact that Aristotle's definition of time includes the term "motion," and because motion cannot be understood independently of temporal notions, the definition cannot possibly succeed in revealing the essential nature of time – rather, it simply presupposes time. I think that it is now apparent that this charge of circularity is mistaken, for Aristotle does not endorse any version of the at-at theory of motion. His definition of motion is couched in terms of substances and telic properties, not in terms of temporally variant property instantiation. So he is in fact at liberty to define time in terms of motion without thereby falling into circularity.

There is, of course, the residual worry about the fact that "before" and "after" figure in Aristotle's definition of time, and the burden of the following chapter is to address that issue. In wrapping up this chapter, however, I wish to reply to the *tu quoque* objection I postponed from the preceding chapter against my reconstruction of the opening argument of

[38] πέφυκε δὲ ἐκ τῶν γνωριμωτέρων ἡμῖν ἡ ὁδὸς καὶ σαφεστέρων ἐπὶ τὰ σαφέστερα τῇ φύσει καὶ γνωριμώτερα· οὐ γὰρ ταὐτὰ ἡμῖν τε γνώριμα καὶ ἁπλῶς. διόπερ ἀνάγκη τὸν τρόπον τοῦτον προάγειν ἐκ τῶν ἀσαφεστέρων μὲν τῇ φύσει ἡμῖν δὲ σαφεστέρων ἐπὶ τὰ σαφέστερα τῇ φύσει καὶ γνωριμώτερα. Cf. *An. Post.* 1. 2, 72b33.

Physics IV.11. Doing so will give us further insight into Aristotle's method and mindset in the construction of his temporal theory.

Coope takes Aristotle to be reasoning about time from the standpoint of common sense. On her reading, his project is philosophically naïve, since it proceeds in *Physics* IV.11 by way of discovering certain underlying assumptions that we all allegedly make, but of which we were previously unaware. I reject this reading, because I do not think that Aristotle's arguments are intended to help him discover what time is; I believe that he comes to the table with a fully developed view in hand and that his arguments are intended to convince the reader that his definition is correct. But if Aristotle's arguments are polemical rather than probative, what should prevent him from deploying other elements of his philosophical arsenal? Specifically, why shouldn't Aristotle exploit the account of motion that he gives in the preceding book of the *Physics*, the correct interpretation of which, as I have just explained, is far from naïve?

Nothing should prevent him from doing so. Indeed, it is rather obvious that he *must* have this account in mind if his definition of time is to avoid circularity. Coope herself notes that Aristotle's peculiar, non-temporal definition of motion is what permits him to define time noncircularly in terms of motion.[39] Aristotle unveils his definition of time very shortly after rehearsing the arguments that open IV.11, and yet he nowhere indicates that he is abandoning some intuitive or common-sense conception of motion in favor of his own, theoretically sophisticated definition. We ought to conclude, then, that he has had the latter in mind all along. Thus, Aristotle's account of time is very far from being naïve, and not only does he not (*pace* Coope) implicitly argue to the conclusion that motion could not exist without time – a claim he rejects in IV.14 without any argument – there is simply no *need* for him to argue to the contrary conclusion. For the conception of motion he employs in the development of his theory of time is such that temporal notions do not figure in it anyway.

This line of reasoning fortifies my critique of Coope's interpretation of the argument in IV.11, but it also makes a more general point about the kind of exegesis in which we are engaged. Aristotle is wont to play the part of common-sense philosopher, and pretheoretical intuitions certainly do have a role in his thinking. But it is a grave mistake to believe that his theoretical arguments are assembled in the ordinary light of common

[39] Although our respective interpretations differ on several points of importance, Coope deserves credit for highlighting this fact and exploiting it in her argument (2005). See p. 9 and p. 363 n.12.

experience. He freely employs language that has a home in the commerce of regular discourse, but more often than not, his use of that language is vastly more sophisticated than it might seem, and commonplace terms are laden with theoretical significance. So if we ourselves approach his theory of time naïvely, there is a genuine risk that we will not appreciate the substance, let alone the subtlety, of the view.

CHAPTER 5

"The before and after in motion"

Aristotle's non-temporal conception of motion saves his definition of time as "a number of motion with respect to the before and after" from circularity, at least insofar as "motion" appears in the definiens. But the definition will nonetheless fail unless "before" and "after" can also be explicated independently of temporal notions. And they can be.

It goes without saying that "before" and "after" do indeed denote familiar, reciprocal temporal relations: Armstrong's moon landing is (in some timeless sense of "is") before Nixon's death – and the latter is after the former – because the landing occurred at a date earlier than the death. But as this illustration makes amply clear, every relation requires relata, and so I shall speak of the before and after in motion, as well as in time, in terms of the entities that function as the relata for each respective relation, whatever those relata might turn out to be.[1] On the hylomorphic interpretation of Aristotle's theory of time, the temporal entities denoted by "before" and "after" as used in their temporal sense must have some variety of kinetic entities as their matter. Moreover, the existence of such kinetic entities provides a proper referent for "before" and "after" used in a distinctly kinetic sense. My contention is that Aristotle uses the terms in precisely this sense in his definition of time, and that the definition is therefore not circular in any respect.

If the kinetic before and after are to function as the matter for the before and after in time, it seems as though a sufficiently high degree of similarity between the two kinds of entities should obtain. There are two distinct issues to be addressed in connection with this similarity relation.

The first issue concerns the intrinsic character of the entities themselves, and most importantly, their *size*. On the face of it, the before and after in time are either temporal intervals or extensionless moments;

[1] I am aware that "entities" is often used in its traditional sense as referring to ontologically self-sufficient objects (such as Aristotle's primary substances). My use of it in this context doesn't presuppose that traditional sense; I use it simply for convenience.

whichever is the case, we should expect their kinetic material constituents to be of similar character, either kinetic intervals or kinetic "points." My treatment of this issue looks to Simplicius' commentary on the *Physics*, in which he argues that they are intervals. I offer reasons for rejecting Simplicius' reading and lobby for an interpretation according to which the kinetic before and after are entities with zero extension.

The second issue concerns the ordering relation between related pairs of befores and afters. Determining temporal priority and posteriority is a perfectly unproblematic matter if we know the relevant historical facts. But by virtue of what do the kinetic before and after stand in relations of kinetic priority and posteriority? I argue that these relations are to be found in Aristotle's account of motion itself, and that the directionality of time derives from the intrinsic direction of its material constituent, namely the kinetic continuum. My argument here responds to the work of Denis Corish, who articulates very clearly the most potent version of this complaint against Aristotle's definition of time.

5.1 THE SIZE PROBLEM

On the hylomorphic interpretation of Aristotle's temporal theory, motion serves as the matter of time. We shall have to examine the textual evidence for this interpretation, beginning with a remark he makes in IV.11:

> Now the before and after [*proteron kai husteron*] is primarily in place. There, to be sure, it is by arrangement – but since the before and after is in magnitude, it must also be in motion, analogously with the first. Furthermore, the before and after is also in time, since the one always follows [*akolouthein aei*] the other. (219a14–19)[2]

This passage reveals two crucial facts about the kinetic before and after. First, as the abundant use of the singular form of verbs (and in the English translation, the singular impersonal pronoun "it") shows, "before" and "after" do not denote two essentially different kinds of entity; rather, the entire phrase "before and after" denotes a single type whose tokens stand in relations of posteriority and priority with their fellow tokens. That is, because priority and posteriority are relations, there isn't anything about one of the tokens considered just in itself by virtue of which it is a before rather than an after. A single token might well be a before relative to

[2] τὸ δὴ πρότερον καὶ ὕστερον ἐν τόπῳ πρῶτόν ἐστιν. ἐνταῦθα μὲν δὴ τῇ θέσει· ἐπεὶ δ' ἐν τῷ μεγέθει ἔστι τὸ πρότερον καὶ ὕστερον, ἀνάλογον τοῖς ἐκεῖ. ἀλλὰ μὴν καὶ ἐν χρόνῳ ἔστιν τὸ πρότερον καὶ ὕστερον διὰ τὸ ἀκολουθεῖν ἀεὶ θατέρῳ θάτερον αὐτῶν.

another token and yet an after relative to some third. So for the purposes of addressing the question of their size, I shall in this section use the somewhat awkward expression "before-and-after" as a (grammatically) singular term to refer indiscriminately to any such token, since there is no need to consider them in pairs for the purposes of addressing the size problem.

The second important point made in this passage is that the kinetic before-and-after is imagined somehow to "follow" (*akolouthei*) the spatial before-and-after. This fact about Aristotle's view is a specific instance of a more general feature of the account, namely that motion *itself* "follows" spatial magnitude, and time "follows" motion:

> Since the *kinoumenon* moves from something to something, and since every magnitude is continuous, motion follows [*akolouthei*] magnitude; for it's because magnitude is continuous that motion is also continuous, and it's because motion is that time is. For whatever size the motion is, the time always seems to have been just as much. (219a10–14)[3]

The details of this compact argument needn't concern us at the moment. What's noteworthy is that the continuity of motion is said to "follow" from the continuity of magnitude, and that the continuity of time "follows" from that of motion. Given that Aristotle is here attempting to explain the continuity of time and motion in terms of the continuity of spatial magnitude, the most natural way of understanding the notion of "following" is in terms of some form of metaphysical dependence, and more specifically in this context, as structural dependence. That this is a promising reading of "*akolouthei*" is confirmed by its use in various other treatises independently of the trio presently under examination.[4] Moreover, Aristotle elsewhere maintains (without using the language

[3] ἐπεὶ δὲ τὸ κινούμενον κινεῖται ἔκ τινος εἴς τι καὶ πᾶν μέγεθος συνεχές, ἀκολουθεῖ τῷ μεγέθει ἡ κίνησις· διὰ γὰρ τὸ μέγεθος εἶναι συνεχὲς καὶ ἡ κίνησίς ἐστιν συνεχής, διὰ δὲ τὴν κίνησιν ὁ χρόνος· ὅση γὰρ ἡ κίνησις, τοσοῦτος καὶ ὁ χρόνος αἰεὶ δεκεῖ γεγονέναι.

[4] For examples of contexts in which Aristotle uses ἀκολουθεῖ in the sense of logical or metaphysical dependence, see *GA* 778b5, *HA* 590a16, *An. Pr.* 43b4, *Met.* 981a27, and especially 1003b23 and 1018a36, where he says that *being* (τὸ ὄν) and *one* (τὸ ἕν) are symmetrically so related. For examples of cases where the expression is used in the sense of purely logical dependence, see *An. Pr.* 32a24 and *Top.* 114a4. For examples where it is used in the sense of metaphysical or physical dependence, see *Meteor.* 340b17, *GA* 764b25, 766a29, and 784a27. Also cf. Hussey's (1983, 144) reading of ἀκολουθεῖ in terms of "structure-preserving mapping": "In fact, it is true in general, for each particular aspect of structure [sc. of magnitude, motion, and time], that, while the definition of natural mapping determines the *meaning* of Aristotle's claim that the mapping is structure-preserving, the *justification* of the claim requires further argument which Aristotle does not here provide."

of following) that time, motion, and magnitude stand in progressive dependence relations with respect to being infinite (*apeiron*),[5] and even with respect to simply being a quantity (*poson*).[6]

I won't argue in the abstract for a hylomorphic interpretation of the *following* relation in the relevant context. Instead, I shall simply continue to develop the interpretation on its own terms, focusing next on Simplicius' answer to the question of the size of the kinetic before-and-after.

5.1.1 *Simplicius' account*

Simplicius' commentary on the *Physics* is highly instructive on many topics. On the subject of the kinetic before-and-after, however, Simplicius' grasp of Aristotle's view seems uncharacteristically confused. In the passage below, Simplicius raises the size problem and offers a solution that he himself regards as rather peculiar. I first quote at length the portion of the *Physics* passage upon which Simplicius is commenting and then the commentary itself. Immediately before unveiling his official definition of time, Aristotle writes:

> Now the before and after is primarily in place. There, to be sure, it is by arrangement – but since the before and after is in magnitude, it must also be in motion, analogously with the first. Furthermore, the before and after is also in time, since the one always follows the other. The before and after in motion is, with respect to its substrate, identical with motion; with respect to its being, however, it is different, i.e. it is not identical with motion. Additionally, we also come to know time whenever we mark off [*horisômen*] motion, marking it off by means of the before and after; and we say that time has passed whenever we have a perception [*aisthêsin*] of the before and after in motion. We mark off motion by taking them to be different from one another, and different from another thing between them. For whenever we conceive of the outermost points [*akra*] as different from the middle, and the soul says that there are two nows [*ta nun*], the one before, the other after, that's when, and that's what, we say time is; for what's marked off by nows is thought to be time – let this much be granted. So whenever we perceive the now as one, and neither as before or after in motion,

[5] *Phys.* III.7: "The infinite is not the same in magnitude and motion and time as a single nature, but that which is posterior is said to be infinite by virtue of that which is prior; e.g. on the one hand, motion is because the magnitude along which the thing is moved, alters, or grows, and time is because motion is" (τὸ δ' ἄπειρον οὐ ταὐτὸν ἐν μεγέθει καὶ κινέσει καὶ χρόνῳ, ὡς μία τις φύσις, ἀλλὰ τό ὕστερον λέγεται κατὰ τὸ πρότερον, οἷον κίνησις μὲν ὅτι τὸ μέγεθος ἐφ' οὗ κινεῖται ἢ ἀλλοιοῦται ἤ αὐξάνεται, ὁ χρόνος δὲ διὰ τὴν κίνησιν) (207b21–5).

[6] *Met.* Δ.13: "[F]or because it [the space through which something is moved] is a quantity, motion is also a quantity, and because this is a quantity time is so" (τῷ γὰρ ποσὸν εἶναι ἐκεῖνο καὶ ἡ κίνησις ποσή, ὁ δὲ χρόνος τῷ ταύτην) (1020a31–2).

or as the same but relative to something else which is before or after, it seems to us that no time has passed because it also seems that there is no motion, either. But whenever we do perceive the before and after, that's when we say that there is time. (219a14–219b1)[7]

Simplicius comments:

Having thus separated the before and after in motion from motion, it remains for [Aristotle] to investigate the essence of being before and after, through which they differ from motion. It is like the case of a line: so long as we think of it as one and continuous, we do not think of the portion of it that is before in position, and that which is after, but when we divide it and distinguish points, we at once make one before and one after and treat the extremes as other than the middle of the line that they include. It is the same with motion: we do not think of the before and after in it except when we divide and distinguish it into, perhaps, two stages that we *perceive as* different things. How, then, do we distinguish? Presumably not by a point, since motion has no position, but in some other way. For I call "before" the motion that I have undergone when, perhaps, walking, "after" the motion that I shall undergo. So the distinction is made by one having passed and the other being future. But the past and the future are time. So time is the same thing as the before and after in motion, when they are distinguished. They are distinguished when the nows strike the soul as two, one as the beginning of motion, the other as the end. For then we at once conceive of the time between the nows as different from them, and often we may take one now as the end of what has passed and as the beginning of the future and in that way again we gain a conception of time as both before and after.[8]

There are several noteworthy features of Simplicius' commentary here. First, Aristotle speaks both of distinguishing (*horizômen*) and perceiving (*aisthanômetha*) the before-and-after in motion, whereas Simplicius speaks of distinguishing them and perceiving them as being different things, but

[7] τὸ δὴ πρότερον καὶ ὕστερον ἐν τόπῳ πρῶτόν ἐστιν. ἐνταῦθα μὲν δὴ τῇ θέσει· ἐπεὶ δ' ἐν τῷ μεγέθει ἔστι τὸ πρότερον καὶ ὕστερον, ἀνάγκη καὶ ἐν κινήσει εἶναι τὸ πρότερον καὶ ὕστερον, ἀνάλογον τοῖς ἐκεῖ. ἀλλὰ μὴν καὶ ἐν χρὸνῳ ἔστιν τὸ πρότερον καὶ ὕστερον διὰ τὸ ἀκολουθεῖν ἀεὶ θατέρῳ θάτερον αὐτῶν. ἔστι δὲ τὸ πρότερον καὶ ὕστερον ἐν τῇ κινήσει ὃ μέν ποτε ὂν κίνησις [ἐστιν]· τὸ μέντοι εἶναι αὐτῷ ἕτερον καὶ οὐ κίνησις. ἀλλὰ μὴν καὶ τὸν χρόνον γε γνωρίζομεν ὅταν ὁρίσωμεν τὴν κίνησιν, τῷ πρότερον καὶ ὕστερον ὁρίζοντες· καὶ τότε φαμὲν γεγονέναι χρόνον, ὅτασν τοῦ προτέρου καὶ ὑσέρου ἐν τῇ κινήσει αἴσθησιν λάβωμεν. ὅρίζομεν δὲ τῷ ἄλλο καὶ ἄλλο ὑπολαβεῖν αὐτά, καὶ μεταξύ τι αὐτῶν ἕτερον· ὅταν γὰρ ἕτερα τὰ ἄκρα τοῦ μέσου νοήσωμεν, καὶ δύο εἴπῃ ἡ ψυχὴ τὰ νῦν, ιὸ μὲν πρότερον τὸ δ' ὕστερον, τότε καὶ τοῦτό φαμεν εἶναι χρόνον· τὸ γὰρ ὁριζόμενον τῷ νῦν χρόνος εἶναι δοκεῖ· καὶ ὑποκείσθω. ὅταν μὲν οὖν ὡς ἓν τὸ νῦν αἰσθανώμεθα, καὶ μὴ ἤτοι ὡς πρότερον καὶ ὕστερον ἐν τῇ κινήσει ἢ ὡς τὸ αὐτὸ μὲν προτέρου δὲ καὶ ὑστέρου τινός, οὐ δοκεῖ χρόνος γεγονέναι οὐδείς, ὅτι οὐδὲ κίνησις. ὅταν δὲ τὸ πρότερον καὶ ὕστερον, τότε λέγομεν χρόνον· τοῦτο γάρ ἐστιν ὁ χρόνος, ἀριθμὸς κινήσεως κατὰ τὸ πρότερον καὶ ὕστερον.

[8] Urmson (1992), 712, 28–713, 13. Emphasis added.

never of perceiving them as such. And he is right to refrain from putting it as Aristotle does: since Simplicius conceives of the kinetic before and the kinetic after as being past and future stages of motion, respectively, it is impossible that anyone should be able to perceive the after in motion *as after*. This follows from the obvious fact that no one can perceive movements that will occur in the future: one may anticipate future movements (or stages of a movement), but one can perceive these only when they are present and no longer future. Of course, similar considerations apply to stages of a movement that are past. One can *recall* the movement, but one can't very well perceive it.[9]

In fairness to Simplicius, Aristotle does speak of the before and after in time as being past and future events in the *Metaphysics*:

> [Some] things are prior in time; some by being further from the present, i.e. in the case of past events (for the Trojan war is prior to the Persian, because it is further from the present), others by being nearer the present, i.e. in the case of future events (for the Nemean games are prior to the Pythian, if we treat the present as beginning and first point, because they are nearer the present). (1018b14–19)[10]

It is not entirely obvious to me that this passage deserves much weight in explicating the kinetic before-and-after, since that notion is functioning in a rather technical capacity in Book IV of the *Physics*, and the *Metaphysics* passage aims merely at distinguishing different senses of "before" and "after."

A second and more substantive problem with Simplicius' treatment of the passage is that he hasn't provided any non-temporal criterion by which we can distinguish the kinetic before from the kinetic after, as he construes them. We are left to believe that the before and the after in motion are to be distinguished by making reference to times. But then there would be available no non-temporal sense of "before" or "after" to figure in Aristotle's definition of time, therefore rendering it objectionably circular after all. As a matter of exegetical principle, we ought to resist attributing to Aristotle a plainly circular definition unless we are driven to do so by overwhelming and unequivocal textual evidence. Themistius, another ancient commentator, abides by this principle in his treatment

[9] Cf. *Mem.* 449b10 ff. The distinction between perception, properly speaking, on the one hand, and memory and anticipation, on the other, will be taken up as a central issue in Chapter 9.

[10] τὰ δὲ κατὰ χρόνον (τὰ μὲν γὰρ τῷ πορρώτερον τοῦ νῦν, οἷον ἐπὶ τῶν γενομένων, πρότερον γὰρ τὰ Τρωικὰ τῶν Μηδικῶν ὅτι πορρώτερον ἀπέχει τοῦ νῦν· τὰ δὲ τῷ ἐγγύτερον τοῦ νῦν, οἷον ἐπὶ τῶν μελλόντων, πρότερον γὰρ Νέμεα Πυθίων ὅτι ἐγγύτερον τοῦ νῦν τῷ νῦν ὡς ἀρχῇ καὶ πρότῳ χρησαμένων)·

of the phrase. In his response to Galen's reading of the definition as "So time is the number of change in time," Themistius says:

One must know that the before and after in change are not respectively before and after on account of time but rather themselves create the before and after in time.[11]

Though the details of his interpretation are lost, Themistius seems to be on the right track. If Aristotle's definition is to avoid circularity, it cannot include any overtly temporal expressions.

But third, there is reason to reject Simplicius' characterization of kinetic befores-and-afters in terms of stages of movements. Aristotle says, we should recall, that "The before and after in motion is, with respect to its substrate, identical with motion; with respect to its being, however, it is different." The meaning of this statement is not fully obvious, but it does make it clear that Aristotle does not identify the before-and-after in motion with motion itself. Indeed, it is hard to fathom why Simplicius would have opted for the "stage" interpretation of kinetic befores-and-afters, since in the passage under examination Aristotle links them intimately with outermost limits (*akra*) and pairs of nows (*duo … ta nun*). Simplicius does in fact consider the possibility of a point-like entity in motion, but rather than interpret the kinetic before-and-after to be this kind of thing, he laments that fact that Aristotle doesn't provide us with anything like it, which spoils the desired symmetry between magnitude, motion, and time:

Secondly, I want to enquire, if there are these three, magnitude, motion and time, mutually dependent as he says, magnitude being limited by points, time by the now, what is analogous to these, I mean the point and the now, in the case of motion? For there should be some other limit to motion and not the moving thing. For what relationship has this to the now and the point?[12]

Simplicius is right to compare Aristotle's now to a point, since it is made perfectly clear in *Physics* VI.3 that the now, properly so called, is indivisible (233b32–5).[13] I shall explain later just how the moving thing plays a role in the kinetic before-and-after, but for now I only wish to record the fact that the supposed symmetry (conceived in terms of following) between motion and magnitude on thc one hand, and motion and time on the

[11] Urmson (1992), 127.

[12] Urmson (1992), 725, 4.

[13] "Necessarily, too, the now (not spoken of derivatively, but in its own right and primarily) is indivisible and is present in all time" (Ἀνάγκη δὲ καὶ τὸ μὴ καθ' ἕτερον ἀλλὰ καθ' αὑτὸ καὶ πρῶτον λεγόμενον ἀδιαίρετον εἶναι, καὶ ἐν ἅπαντι τὸ τοιοῦτο χρόνῳ ἐνυπάρχειν).

other, could be sustained if the kinetic before-and-after were something with zero extension.

As one final bit of evidence I offer the brief but telling remark made at 220a18–20: "Further, it is also clear that the now is no part of time, nor is the division [*diairesis*] a part of the movement, just as the point is not part of the line."[14] Here we have a specific instance of the magnitude–motion–time trio in which points and nows are compared to a "division" in motion. The division in question is a kinetic before-and-after, and it is clearly conceived here as a *cut* in the movement, something with zero extension.[15]

I believe that the cumulative result of these considerations is quite clear: if a coherent account of the before-and-after in motion can be developed according to which it is point-like, then that account is to be preferred over any of its "stage" interpretation rivals. So I shall undertake the development of just such an account and herewith dispatch the awkward talk of the kinetic before-and-after in favor of the more informative (and more elegant) appellation "kinetic cuts." This reveals, but certainly does not fully justify, my solution to the size problem, and the question of the kinetic continuum's directionality also remains unanswered.[16] That is, by virtue of what is it the case that one kinetic cut is before and another after? This is the more challenging of the two problems, and answering it will provide a fuller justification of my solution to the size problem.

5.2 THE DIRECTIONALITY PROBLEM

Aristotle's remark in *Physics* IV.14 that motion could exist in the absence of time is evidence that he conceives the essential nature of motion in temporal-free terms. Perhaps this explains why greater attention has been paid to the alleged circularity introduced by his use of "before" and "after" in his definition of time. For even if motion can be understood independently of time, how can the relations of priority and posteriority, as they relate to motion, properly be understood independently of time? This, it turns out, is a central obstacle to appreciating the nature of Aristotle's view, and it is one over which better minds than mine have stumbled.

14 καὶ ἔτι φανερὸν ὅτι οὐδὲν μύριον τὸ νῦν τοῦ χρόνου, οὐδ' ἡ διαίρεσις τῆς κινήσεως, ὥσπερ οὐδ' ἡ στιγμὴ τῆς γραμμῆς·

15 Hardie and Gaye (Barnes 1984) translate διαίρεσις as "section," but this cannot be correct given that Aristotle immediately says that lines are parts of a line, and every line segment (or section) is itself a line. Coope (2005), 65 ff., also conceives the before and after in motion as stages.

16 By "directionality" I mean its possessing a direction at all, not what its direction happens to be.

Julia Annas, G. E. L. Owen, W. D. Ross, and scores of other commentators have dismissed Aristotle's temporal theory on just these grounds. But they are wrong to have done so. For Aristotle has available the theoretical resources to specify a uniquely kinetic sense of "before" and "after," and those terms are used precisely in this sense in his definition of time.

As a way of motivating my account of kinetic priority and posteriority, I shall focus on Denis Corish's articulation of the objection, for he presents it in a very pointed and perspicuous form. Elements of his version of the objection are voiced by the philosophers named above, and I shall mention them again in the course of examining Corish's case. Once the error of the objection is identified, the shape of the solution to the problem of the kinetic continuum's directionality is surprisingly simple and the resources for its realization are readily available.

5.2.1 *Corish's account*

Corish claims that Aristotle's attempt to derive temporal entities from spatial entities fails on account of the fact that there are no correlative kinetic entities which can be explicated independently of temporal concepts. He articulates his principal objection to Aristotle's view in the following passage:

> I have said that it is not clear how ... the [before] and [after] of [motion], which, according to [Aristotle], correspond, it seems, simply to spatial positions, become the [before] and [after] of time, by being associated with [now]. We can assign different nows to spatial positions, so that a moving something can be described as being now at this position, now at that – but this is to see movement in terms of space and time, and not time in terms of space and movement, as Aristotle does.[17]

Here Corish alleges that kinetic cuts are either nothing over and above spatial locations, or are a function of spatial locations and dates. These alternatives, he says, give rise to a dilemma:

> For either the pretemporal order is merely spatial or else it is independent, an independent interpretation, as we should say, of the continuum. But it cannot really be the latter, as we have seen, since there are no independent positions of movement; they are spatio-temporal entities. And if the pretemporal order is merely spatial, then, as we have also seen, there is no way of deriving the temporal order from them.[18]

[17] Corish (1976), 246. The bracketed translations here are supplied for the sake of readability (Corish retains the Greek).

[18] Corish (1976), 250–1.

It seems quite evident to me that Aristotle is prepared to grasp the second horn of this dilemma, since he gives us an account of the kinetic continuum that, when properly understood, provides an "independent interpretation of the continuum." Corish's insensitivity to this response is most likely the result of the fact that he is employing a conception of motion that is decidedly un-Aristotelian:

If we wish to speak of a starting *position* of movement as such, as opposed to the spatial or the temporal coordinates, we must speak of it as a spatio-temporal entity, a point-instant – a point on a plane for which the coordinate axes are space and time.[19]

The idea here can be generalized to apply to any kinetic cut, so that its specification will include reference to a spatial location as well as reference to a date. The reference to time, Corish thinks, is needed to determine which member in a pair of kinetic cuts is the "before" and which is the "after." But since the before and after in time are supposed somehow to be *derived from* the before and after in motion, no specification of a kinetic cut may make reference to time without thereby undermining the prospects of the desired derivation. Hence, on Corish's interpretation, Aristotle's project is doomed to fail.

Corish's objection is echoed by others who have addressed the directionality problem. Consider Julia Annas' summary dismissal of Aristotle's claim that time follows motion:

It would in fact be an error to make time logically derivative from motion, because motion or change already involves time. Aristotle recognizes this at 222b30–223a15: all motion is (relatively) fast or slow, and this involves the notion of covering a distance in more or less time. But he does not remain sufficiently aware of this, or he would have suspected a covert circularity in the scheme in which before and after in time is derived from before and after in motion and this in turn from a primary before and after in space.[20]

The passage Annas cites does not say, nor does it suggest, that all motion is *essentially* fast or slow, which is the premise required for Annas' argument to go through. Consider the following: while it may be true (let us suppose) that every medium-sized physical object is colored, there is nothing suspect in developing a theory of color according to which its existence is partly dependent upon the existence of medium-sized physical objects. Hence, the passage Annas cites cannot be offered as compelling evidence that Aristotle believes that "motion or change already involves time."

[19] Corish (1976), 250. [20] Annas (1975), n. 11.

Annas' charge of covert circularity is expressed very compactly, which is quite typical for this particular line of objection. Consider another version of the objection offered by G. E. L. Owen:

> But can this direction [of motion] be derived from the spatial before-and-after we have just defined, without importing just the temporal priority we meant to explain? Evidently not. We might define a direction *abc for [the spatial magnitude], by saying that ac contains ab but ab does not contain ac; but of course we could on the same terms define the direction *cba ... And if we try to sharpen the condition by specifying where on the line the movement begins or ends, our explanation of temporal order becomes immediately circular.[21]

Because Aristotle conceives the before and after in magnitude as a pair of spatial positions whose ordering is purely conventional, as Owen says, there's nothing about either member of a pair considered as such that makes one before and the other after.[22] But the temporal before and after, conceived as a pair of distinct instants, are intrinsically ordered, and there is a real difference between the past and the future.[23] So to claim that the before and after in time are somehow derived from the before and after in magnitude immediately raises the question of whether the before and after in motion are intrinsically or conventionally ordered. If intrinsically, then something magical seems to have taken place in the derivation of motion from magnitude. If conventionally, then the derivation of time from motion would have to be equally magical. Hence, whichever of the two ways the derivation is supposed to run, Aristotle's account seems to be trying to spin hay into gold.

So Corish is in good company when he claims that Aristotle's account of time is viciously circular. But neither he nor any of his fellow critics is thinking about motion in the way that Aristotle does. As we saw above, Aristotle rejects the at-at conception of motion. Indeed, we find surprisingly little use of temporal concepts in Aristotle's discussion of motion in *Physics* III.

One way of bringing out the error in Corish's analysis is to recall the fact that every movement is a movement *of something*. "There is no motion apart from actual things," Aristotle tells us in *Physics* III.1 (200b32–3).[24] More proximate to his definition of time, he says at 219b23–5, "For it is by the moving thing that we become acquainted with the before and after in

[21] Owen (1976), 24–5.
[22] Recall 219a14, quoted above. Also cf. *Met.* Δ.11, 1028b26–9.
[23] Consider Aristotle's discussion of the difference between past (or present) and future contingent sentences in *Int.* 9, where he offers his famous sea-battle example.
[24] οὐκ ἔστι δὲ κίνησις παρὰ τὰ πράγματα.

motion."[25] To provide a satisfactory account of an "independent" kinetic continuum, then, we need somehow to incorporate the subject of motion. I put "independent" in shudder-quotes because the analysis of kinetic positions I attribute to Aristotle includes references to spatial locations and is thereby very much dependent upon them. However, this feature of my solution to the directionality problem ought to be viewed as a virtue rather than a vice, Corish's demand for an "independent interpretation of the continuum" notwithstanding. What is crucial for a serviceable analysis of kinetic cuts, again, is not that it make no reference to spatial locations, but rather that it make use of them in some substantive way, and that it not make reference to dates. On the hylomorphic interpretation of Aristotle's account of time, temporal order is supposed to derive from kinetic order, which in turn is supposed to derive from spatial order. This, then, is our present challenge.

5.3 KINETIC CUTS DEFINED

My account of kinetic cuts makes use of the interpretation of Aristotelian motion presented in Chapter 4, according to which motion is the actuality of a peculiar accidental compound, a substance just insofar as it possesses a telic property. Aristotle employs the kind-term *kinoumenon* for such telic property compounds, much as he uses the singular term "cloak" in *Metaphysics* Z.4 to refer to the accidental compound to which he also refers by the definite description "the pale man."

My task of defining kinetic cuts will be greatly simplified by introducing a bit of shorthand. First, let us recall Aristotle's commitment to the claim that every particular movement involves a *hupokeimenon* and a pair of contraries: every change, he insists, is a change from some initial state toward some end or *telos*. So "*kinoumenon*" is in fact a variety of determinable whose determinate instances include the *telos* of the movement. Diares' being a *kinoumenon* logically requires that he be some *F*-type *kinoumenon*, where '*F*' specifies the *telos* of the movement. So in our stock example, Diares would be an *on-Crete kinoumenon*. I shall abbreviate this description as follows:

kin(on Crete)

This description could be used to refer obliquely to Diares himself, independently of any considerations about his present or future location. But

[25] τῷ γὰρ φερομένῳ γνωρίζομεν τὸ πρότερον καὶ ὕστερον ἐν κινήσει.

the function of the "*qua* such" qualification in Aristotle's definition of motion is to restrict the reference to the *kinoumenon* as such. I employ the shriek ("!") to formally represent definite description functioning to refer strictly to the compound entity. So to represent this variety of reference to *the on-Crete kinoumenon*, we shall write:

!*kin*(on Crete)

With this bit of notational machinery, defining kinetic cuts is a surprisingly simple matter. Considering again Diares' voyage from Rhodes to Crete, the spatial locations along which his movement takes place could be plotted on a map, and the dates of occupation for each of those locations could also be specified, if we happen to have an accurate itinerary. But the kinetic cuts included in the movement are not (*pace* Corish and his fellow objectors) functions over the members of these two sets. Rather, they exhibit the logical structure of the following:

<!*kin*(on Crete), immediately south of Kassos>

A kinetic cut is specified by an ordered pair whose elements are an *F*-type *kinoumenon* considered as such, and a spatial location on the *F*-bounded continuum at which the *kinoumenon* in question is actually located. Thus, as this example should make clear, I take kinetic cuts to be *states* of *kinoumena* specified in terms of spatial locations.[26] Of course, "south of Kassos" isn't nearly precise enough to give us a genuinely point-like kinetic entity. But I hope that the larger point is not missed for that technicality: because *kinêsis* is essentially telic, on Aristotle's view, the direction of the kinetic continuum is not derived from time, but rather from the essential nature of motion itself.

Note that by the above explanation the following does not pick out any kinetic cut:

<!*kin*(on Crete), possessing a mass of exactly sixty kilograms>

The first member of the pair limits the range of predicates that could legitimately function as the second member of the pair to spatial locations, and more specifically to spatial locations lying on the *kinoumenon*'s trajectory. Not just any state of a *kinoumenon* properly counts as a kinetic cut – the state must be one specifiable within the continuum of states specified in the proper description of the *kinoumenon*.

[26] Again, I frame the interpretation in terms of locomotion; other species of *kinêsis* may be handled in a similar fashion.

So to illustrate the relation of kinetic priority and posteriority, imagine that I wish to return a book to the library, which happens to be north of my office in Lincoln Hall. As I sit in my office, I possess the plasticity property *being possibly at book drop*. This second-order modal property is a genuine feature that I possess. But although I am the substance that will serve as the *hupokeimenon* of the movement in question, it is not the case that I possess a modal property of the right sort to be a *kinoumenon* – while it is true that I am possibly at the book drop, it is equally true that I am possibly at the pub to the south of my office. *Kinoumena* are essentially partly constituted by a telic property that precludes its possessor from instantiating any correlate telic property. My possibly being at the book drop and my possibly being at the pub are equally present in me as I sit in my office.

But when I quit my office and exit Lincoln Hall to the north, I instantiate a telic property that precludes my instantiating the correlate telic property involving the pub. Described intuitively, I am the material component of an actual *kinoumenon*, one to which we may refer by means of the description "!*kin*(at the book drop)." Now consider two of the kinetic cuts included in my walk to the library:

a: <!*kin*(at the book drop), fifty-one meters from the book drop>
b: <!*kin*(at the book drop), fifty meters from the book drop>

Kinetic priority and posteriority is grounded in spatial inclusion. Specifically, pairs of kinetic cuts whose *kinoumena* are numerically identical stand in such kinetic relations, and the relations in question are fully determined by the spatial intervals constituted by the telic property and the spatial location specified in the ordered pair. More precisely:

Kinetic order: for any pair of related kinetic cuts x and y ($x \neq y$), x is *before* y iff the spatial interval specified in y is included in the spatial interval specified in x; otherwise, x is *after* y.[27]

Thus, in the example above, because the interval [the book drop, fifty meters from the book drop] is included within the spatial interval [the book drop, fifty-one meters from the book drop], this definition yields the result that kinetic cuts (*a*) and (*b*) are respectively the before and after in my movement.

[27] This definition applies to individual movements (where the "related kinetic cuts" share a common *kinoumenon*). I shall indicate in Chapter 11 how such relations can be extended to cover different *kinoumena*, which thereby can serve as the matter for temporal relations between different movements and events.

The virtues of this analysis should be apparent. On this interpretation, individual kinetic cuts are specified in terms of spatial locations, and the ordering relation among them is determined crucially by making reference to such locations. So this interpretation conforms with Aristotle's claim at 219a14–21 that the before and after in motion follow the before and after in magnitude – the elements and the structure of the kinetic continuum derive from the elements and structure of the spatial continuum. But temporal notions figure nowhere in the analysis. It has long been understood that Aristotle conceives motion to be essentially telic, or directional, in nature. Unfortunately, this feature of Aristotle's view of motion has not been sufficiently exploited by those attempting to reconstruct his theory of time. Once Aristotle's non-temporal, telic conception of motion is taken seriously, the charge of circularity leveled by Corish and so many others simply evaporates. The elements and structure of the kinetic continuum can be defined without making reference to temporal entities or temporal ordering relations.

5.4 *KINÊSIS* VS. KINETIC CUTS

The preceding analysis of kinetic cuts deflects the circularity objection, but it is itself vulnerable to a fairly obvious objection. I have argued that Aristotle defines motion as the actuality of a *kinoumenon*, considered as such. I have also argued that a proper understanding of Aristotle's notion of kinetic cuts ("the before and after in motion") requires reference to *kinoumena*. But because the hylomorphic interpretation of Aristotle's temporal theory treats kinetic cuts as the matter of instants in time, it appears as though a consequence of this interpretation is an affirmation of instantaneous motion. But Aristotle explicitly denies the possibility of instantaneous motion in *Physics* VI.3 and later in VI.9, where he addresses Zeno's paradoxes of motion. Therefore, it appears as though the hylomorphic interpretation cannot be correct.

There is a wealth of fine scholarship on this particular issue, two of the more prominent examples belonging to Owen and Broadie.[28] I have no intention of directly engaging that body of literature, for I think that there are several texts that clearly favor the hylomorphic interpretation's treatment of kinetic cuts as the matter of instants of time and also show that Aristotle's position here is consistent with his rejection of instantaneous motion in *Physics* VI. And as it turns out, the response to this

[28] Owen (1975c), Broadie (1983).

objection enhances the plausibility of the explication of Aristotle's definition of motion provided above.

Let us begin by considering Aristotle's remark at 219a19–21:

> The before and after in motion is identical in substratum [*ho pote on*] with motion yet differs from it in being and is not identical with motion.[29]

Ross maintains that the puzzling phrase "*ho pote on*" (literally "being which ever") in this context is an ellipsis for "*touth ho pote on proteron kai husteron en tê kinêsei estin*" (literally "that, being which ever the before and after in motion is") and initially suggests that it denotes the *hupokeimenon*, or underlying subject, of the movement.[30] But immediately after invoking the *hupokeimenon*, Ross offers the following gloss: "The general sense of the sentence must be 'the before and after in movement is, as regards its subject, movement; but its essence is not movement'."[31] The change of mind is puzzling, and I am not sure how to explain it.

Coope dedicates an entire appendix of her book to the odd phrase in question, noting that it occurs in the corpus in only three places outside of the discussion of time in the *Physics*.[32] She attributes to Philoponous and Simplicius the *hupokeimenon* reading initially endorsed by Ross, but she rejects it on grounds that it is too obvious an interpretation, given the rarity of the phrase.[33] She provides a very nice analysis of the grammatical possibilities, but ultimately endorses a reading of the passage that is precisely the opposite of Ross':

> It must, then, be that *ho pote on X esti* is something that provides the ground for X's being what it is … Aristotle says that that by being which change is is the before and after in change. If I am right about the meaning of this, then he is implying that the before and after in a change provides an explanatory ground for the change's being the change it is.[34]

Whereas Ross takes motion to be the *ho pote on* of the before and after in motion, Coope takes the before and after in motion to be the *ho pote on* of motion. Which of them has it right?

[29] ἔστι δὲ τὸ πρότερον καὶ ὕστερον ἐν τῇ κινήσει ὃ μέν ποτε ὄν κίνησις [ἐστιν]· τὸ μέντοι εἶναι αὐτῷ ἕτερον καὶ οὐ κίνησις. I have opted to quote Hardie and Gaye's translation, which seems to me to preserve the ambiguity in the original. Ross brackets "ἐστιν" to the disapproval of Coope (2005), 176 n. 8 (discussed immediately below), but my alternative reading is neutral on its inclusion.

[30] Ross (1936), 598. [31] Ross (1936), 598.

[32] Coope (2005), 173. The phrase appears twice in *PA* II (649a15–16, 649b24–5) and once in *GC* I.4 (319b3–4).

[33] Coope (2005), 175 n. 7. [34] Coope (2005), 175.

Neither of them does. Aristotle is not identifying either motion or kinetic cuts as the *ho pote on*, or substrate, of the other. Rather, he is claiming that motion and kinetic cuts share a common *ho pote on*, that they both have, in Coope's terms, the same ground of being. This ground of being is a *kinoumenon*. Aristotle defines motion, I argued, as the actuality of a *kinoumenon*, a telic property compound. Kinetic cuts, on the other hand, are defined as *states* of *kinoumena*. If we wish to employ Aristotle's view concerning accidental compounds, we could say that a kinetic cut is one such type of compound whose matter is a *kinoumenon* and whose form is a location property. The states in question are specifiable as an ordered pair whose elements are some *F*-type *kinoumenon* considered as such, and a spatial location on the *F*-bounded continuum at which the *kinoumenon* in question is actually located, such as the following example:

<!*kin*(on Crete), immediately south of Kassos>

What remains to be explained is how the analysis on offer is consistent with and supports Aristotle's claim that motion and kinetic cuts "differ in being." Let us, then, examine another passage in which motion and kinetic cuts appear in connection with the troublemaking phrase:

> [C]orresponding to the point is moving thing [*pheromenon*], by which we come to know the movement and the before and after in it. This is the same substrate [*ho men pote on*] (whether the point is a stone, or something else of that sort), but it is different in account [*tô logô*], just as the sophists hold that Coriscus in the Lyceum and Coriscus in the marketplace are different. And this is different by being in different places. (219b16–22)[35]

The first sentence unambiguously identifies a species of *kinoumenon* (namely, a *pheromenon*) as serving an epistemic role in our recognition of both motion and kinetic cuts. It is capable of serving this function only if its relation to motion and kinetic cuts is both fundamental and analogous in each case. The relation in question is identified in the very next sentence: it is the same *ho pote on*, but it differs in *logos* or account. Now it is a simple matter to see how the *kinoumenon* differs in *logos* for a pair of

[35] καὶ ὁμοίως δὴ τῇ στιγμῇ τὸ φερόμενον, ᾧ τὴν κίνησιν γνωρίζομεν καὶ τὸ πρότερον ἐν αὐτῇ καὶ τὸ ὕστερον. τοῦτο δὲ ὃ μέν ποτε ὂν τὸ ταὐτό (ἢ στιγμὴ γὰρ ἢ λίθος ἤ τι ἄλλο τοιοῦτόν ἐστὶ), τῷ λόγῳ δὲ ἄλλο, ὥσπερ οἱ σοφισταὶ λαμβάνουσιν ἕτερον τὸ Κορίσκον ἐν Λυκείῳ εἶναι καὶ τὸ Κορίσκον ἐν ἀγορᾷ. καὶ τοῦτο δὴ τῷ ἄλλοθι καὶ ἄλλοθι εἶναι ἕτερον· Presumably, Aristotle here refers to a diagram for the purposes of illustration, and the point in question is taken to represent a stone or similar object. I read the parenthesis at 219b19 with Owen and Miller (cf. Miller, 1974, n. 13) in order to render Aristotle's remark here consistent with his claim at *Phys.* 240b8 ff. that dimensionless items are incapable of motion.

kinetic cuts: the second element of the ordered pairs is distinct. Recall the example employed in the preceding section:

a: <!*kin*(at the book drop), fifty-one meters from the book drop>
b: <!*kin*(at the book drop), fifty meters from the book drop>

These are consonant with Aristotle's remark considering the difference between Coriscus in the Lyceum and Coriscus in the marketplace. He characterizes the difference as sophistic, but I take it that the sophists' view was that Coriscus himself is different by virtue of being in different places. Aristotle regards a substance's location as one of its accidental features, so he rejects the sophists' view. But as regards the before and after in motion, location is essential to the identity of any particular kinetic cut, since they are individuated by means of ordered pairs whose second element denotes a specific location. Aristotle is nonetheless still free to claim that the *ho pote on* of the pair of kinetic cuts – that is, the *kinoumenon* – differs in account in a way that is analogous to the way in which Coriscus himself so differs, namely by being in different places. (We ought to note, though, that while there is nothing that prevents Coriscus from being in the Lyceum, it is impossible that the !*kin*(Lyceum) should be in the Lyceum, given the very nature of telic properties.)

The passage under consideration conforms nicely with our account of kinetic cuts, but Aristotle is making the additional claim that the *ho pote on* of motion differs in *logos* from the *ho pote on* of kinetic cuts. To appreciate the relevant difference, we must look to *Physics* VI. Just four chapters before Aristotle recounts Zeno's paradoxes and denies the possibility of instantaneous change, he provides the key to solving our puzzle:

Now the primary time that has reference to the end of the change is something really existent (for a change may be completed, and there is such a thing as an end of change, which we have in fact shown to be indivisible [*adiaireton*] because it is a limit [*peras*]). But that which has reference to the beginning is not existent at all; for there is no such thing as a beginning of change, nor any primary time at which it was changing. (236a10–15)[36]

By "primary time" Aristotle means the minimally inclusive time. So, for example, it might be true that Smith visited the Lincoln Memorial in 2002, but "2002" does not name the primary time of his visit unless it

[36] τὸ μὲν οὖν κατὰ τὸ τέλος τῆς μεταβολῆς πρῶτον λεγόμενον ὑπάρχει τε καὶ ἔστιν (ἐνδέχεται γὰρ ἐπιτελεσθῆναι μεταβολὴν καὶ ἔστι μεταβολῆς τέλος, ὃ δὴ καὶ δέδεικται ἀδιαίρετον ὂν διὰ τὸ πέρας εἶναί)· τὸ δὲ κατὰ τὴν ἀρχὴν ὅλως οὐκ ἔστιν· οὐ γὰρ ἔστιν ἀρχὴ μεταβολῆς, οὐδ' ἐν ᾧ πρώτῳ τοῦ χρόνου μετέβαλλεν. Cf. the argument in VI.6 that anything that is moving has already moved.

exactly fills that year. More probably, the primary time of Smith's visit to the monument was the interval between 11.40 a.m. and 1.15 p.m. on June 14, 2002.[37] Now Aristotle says immediately prior to the passage in question that "the primary time in which that which has changed has changed must be indivisible" (235b32–3).[38] If we take the time at which a changing thing has changed to be the time of the change's completion, this is clearly consistent with his remark in the quoted passage that the final primary time is indivisible (*adiaireton*) and a limit (*peras*). Expressed in the terminology introduced above, Aristotle commits himself in VI.5 to the existence of a final kinetic cut within any given movement.

But the more important part of the passage for our present purposes is his denial that there is any initial primary time in a movement, which we can express otherwise by saying that there is no first kinetic cut within any given movement. The beginning of a movement is, as we should say, an open interval. Expressed in temporal terms, Aristotle believes that there is a last instant of movement – namely, the instant at which the *kinoumenon* arrives at its destination. Moreover, he believes that there is an instant-from-which the movement begins – namely, the last instant of rest. But he explicitly denies that there is any first instant of movement. His view here is consistent with his expressed view that space, motion, and time are all continuous, by virtue of which fact no two indivisibles within any such continuum are in series.[39]

These commitments provide a way for us to distinguish motion from kinetic cuts. Kinetic cuts are (as intuitively described) instantaneous states of *kinoumena*. Because Aristotle denies the existence of a first instant of movement, there is no first kinetic cut within a movement. Therefore, there can be no last kinetic cut within movement unless there is a prior (though not immediately prior) kinetic cut. So if there can be no first, and there can be no last without a prior, no *kinoumenon* can exist just for a single instant: the being of a *kinoumenon* is necessarily extended. But this is precisely what we should have expected Aristotle to say. In response to Zeno he denies that there is motion (or rest) in an instant. On the

[37] Cf. 236b19–22. More on primary times in Chapter 11.

[38] ἐν ᾧ δὲ πρώτῳ μεταβέβληκεν τὸ μεταβεβληκὸς, ἀνάγκη ἄτομον εἶναι. Aristotle's use of the notion of primary time in *Phys.* VI is derivative in the same way that the notion of an indivisible time is derivative. Instants are indivisible, but they are neither themselves times nor parts of time (cf. 220a18–22). Consequently, no single kinetic cut has a primary time. (See section 11.2.) His use of the notion here is a matter of terminological convenience.

[39] Cf. *Phys.* V.3. Note that to be consistent Aristotle must also deny the existence of a first moment of rest.

interpretation I have endorsed, his defining motion as the existence or being of a *kinoumenon* strictly entails that *kinoumena* cannot exist for a mere instant.

I conclude, then, that the interpretations of Aristotle's account of motion and kinetic cuts offered above are consistent with his rejection of instantaneous motion. Indeed, not only are the two interpretations consistent with his expressed view, they explain and justify that view. This fact provides additional reason to suppose that I've got things right: motion is essentially the existence or being ("actuality," Aristotle would say) of a *kinoumenon*, whereas the before and after in motion (kinetic cuts) are states of *kinoumena*. The way in which the *ho pote on* of motion differs from the *ho pote on* of kinetic cuts is precisely analogous to the difference between being Coriscus and Coriscus' being pale.

5.5 SUMMARY OF PART II

We have covered a lot of ground in these three chapters, so let me summarize the major points before moving on.

Chapter 3 was devoted to examining the opening argument in IV.11, where Aristotle begins to formulate his positive account of time. Against Sorabji, Coope, and others, I argued that the best reconstruction of the "No Time without Motion" argument makes it clear that Aristotle regards the pair <time, motion> as being included in the extension of the evident proper feature (EPF) relation. The nature of this relation is such that one cannot perceive the item in question (the second member of the ordered pair) as such without thereby perceiving it as having the feature in question (the first member). I appealed to the *Topics* in making this case and cited other candidate (EPF) pairs: <colored, surface>; <rarity, fire>; <ensouled, living creature>; <anaplestic, liquid>. A more germane pair is <distance, space>, since the basis for its inclusion in (EPF)'s extension is the same as that for <time, motion>: a certain kind of indeterminacy of the second item situates it as matter to the enforming agency of perception, through which the determinacy of the first item of the pair is realized. But more on this in the next chapter.

Soon after presenting the "No Time without Motion" argument, Aristotle defines time as "a number of motion with respect to the before and after." It is quite natural to think that the definition is objectionably circular on grounds that several of the terms in the definiens – specifically, "motion," "before," and "after" – seem to be ineliminably temporal in nature.

I argued that to dismiss the definition as circular for its inclusion of "motion" is to neglect Aristotle's account of motion in Book III of the *Physics*, which is developed in temporal-free terms. Motion, he says, is "the actuality of what is potentially, *qua* such." In Chapter 4 I developed a reading of the definition according to which the unit-phrase "what is potentially" refers to a compound entity whose constituents are (typically) a substance and (always) a telic property. An example of such a compound is Socrates insofar as he has the property *being potentially located at the Stoa*. Aristotle takes to calling this type of compound a "*kinoumenon*." The "*qua* such" qualification in the definition ensures that the phrase "what is potentially" refers strictly to the compound and not obliquely to the substance that happens to possess the property. The *actuality* of such a compound is the actuality, or existence, of a *kinoumenon*. And this is just what the essential nature of motion is, on his view.

A proper understanding of Aristotelian motion provides the means of defusing the circularity objection as it applies to "before" and "after" in his definition of time, and in Chapter 5 I articulated Aristotle's response to the objection. There is clear textual evidence that Aristotle regards "the before and after in motion" as picking out point-like kinetic entities. I call these entities "kinetic cuts," and they are states of *kinoumena*. Two simple examples of kinetic cuts are these:

a: <!*kin*(in Manhattan), in Washington, D.C.>
b: <!*kin*(in Manhattan), in Philadelphia>

As Aristotle conceives motion, it is essentially directional in nature, since one of the compounded elements is a telic property. The essentially telic character of motion is reflected in the first member of an ordered pair that denotes a kinetic cut. And by virtue of the telic nature of motion, kinetic cuts are ordered by the very nature of motion itself, not by time. Kinetic cut (*a*) above is before kinetic cut (*b*), because the spatial interval specified in the latter (namely, the interval [Philadelphia, Manhattan]) is included in the spatial interval specified in the former (namely, [Washington, D.C., Manhattan]). Thus, Aristotle is free to define time in terms of kinetic cuts without thereby introducing any circularity in the definition. This interpretation also conforms well with Aristotle's claim that the before and after in motion "follows" the before and after in place insofar as kinetic cuts are partly individuated by means of spatial locations.

The virtues just mentioned notwithstanding, my interpretation of kinetic cuts might seem to commit Aristotle to instantaneous motion insofar as I regard them as the matter for instants in time. This would be a most unfortunate consequence, since Aristotle unequivocally rejects instantaneous motion in his treatment of Zeno's paradoxes. But it is not a genuine consequence of the interpretation. In fact, there is no danger of mistaking kinetic cuts for movements. Aristotle says quite explicitly that there is no first instant of a movement. This entails the impossibility of any kinetic cut existing discontinuously with other kinetic cuts sharing the same *kinoumenon* as a "substrate." So the actuality of a *kinoumenon* is necessarily extended. Therefore, no kinetic cut can be mistaken for the actuality of a *kinoumenon*, which is what motion is, on Aristotle's view.

PART III

The form of time: perception

CHAPTER 6

Number and perception

Part II of this book was devoted to developing an account of the material side of Aristotle's temporal theory. Previous attempts to reconstruct this theory have foundered on the relation between motion and time, and more specifically, on the question of how temporal priority and posteriority could possibly be derived from the kinetic continuum. Unable to solve these problems to their satisfaction, commentators (not surprisingly) never fully saw their way to addressing the formal side of the theory, or indeed even to recognizing that Aristotle's view is essentially hylomorphic in character. Considered in the light of this exegetical history, explicating the matter of time is of no small significance.

However, material analysis is only one half of any hylomorphic account, and it is in some sense the less important of the two. Aristotle reminds us time and again that it is *form* that makes something the very thing that it is, as he does in *Physics* II.1: "Indeed, it [*morphê* – shape, form] is more the nature of a thing than its matter; for something is more accurately said to be what it is when it is in actuality rather than potentiality" (193b6–8).[1] So while the account of time's *hulê* presented in the previous chapters constitutes an important first step toward the development of a fully proper understanding of Aristotle's theory of time, the more challenging portion of the project lies ahead of us.

My treatment of the form of time is multi-staged. The first stage (here in Chapter 6) takes up Aristotle's inclusion of time in the genus *number*. By carefully examining portions of the central chapters of Book IV of the *Physics* in which Aristotle develops and argues for his view, we shall see how the requisite sense of "number" (*arithmos*) leads us to identify perception (*aisthêsis*) as providing the form of time. To summarize the view very compactly, *time* is motion's susceptibility to division into undetached

[1] καὶ μᾶλλον αὕτη [ἡ μορφή] φύσις τῆς ὕλης· ἕκαστον γὰρ τότε λέγεται ὅταν ἐντελεχείᾳ ᾖ, μᾶλλον ἢ ὅταν δυνάμει.

parts of arbitrary length by percipient minds, *the now* is a kinetic cut insofar as it is individuable by perception, and *a time* is a determinate segment of motion as defined by the individuation of kinetic cuts.

The second stage (Chapter 7) addresses an important objection to my interpretation: if Aristotle is committed to the perceptibility of kinetic cuts, then he is in a fairly straightforward sense committed to the possibility of instantaneous perception, which seems a rather dubious view to hold. I will show that Aristotle accepts not only the possibility of instantaneous perception, but positively affirms the actuality of such. In fact, I'll up the ante: not only does he affirm the actuality of instantaneous perception, he endorses the *necessity* (suitably understood) of instantaneous perception. My case for this reading of Aristotle takes us into the heart of his theory of perception, which is articulated in the *De Anima* and extended in various works in the *Parva Naturalia*.

While the second stage is itself demanding, its successful completion gives rise to another puzzle whose solution constitutes the third stage of the formal analysis. If perception is thought by Aristotle to be instantaneous, as I argue it is, then there are familiar perceptual phenomena which would seem to be inexplicable. One anachronistic example of the sort of phenomenon I have in mind is that of watching a movie.[2] The typical frame-rate for celluloid films shown at movie theaters is thirty-two frames per second. If perception were instantaneous, then we wouldn't perceive the images projected on the screen as moving; rather, we would see a series of discrete snapshots. But we obviously *do* perceive the images as moving, and so Aristotle's commitment to the instantaneity of perception would seem to be demonstrably false.

The solution to this challenge requires understanding the role of *phantasia* within Aristotle's theory of perception. *Phantasia* (usually rendered as "imagination") is a vexed element of Aristotle's psychology. Fortunately, I have no intention to say all that can be said about *phantasia*. My principal aim in the third stage (Chapter 8) is to show that there is ample textual evidence to suggest that the kind of phenomenon described above is one with which Aristotle was familiar and whose occurrence he explains by appealing to *phantasia*.

My invoking *phantasia* as sketched above sets the stage for the fourth stage of my explication of the form of time (Chapter 9). In *De Anima* II.6 Aristotle identifies motion as one of the "common perceptibles," with rest, number, figure, and magnitude rounding out the group. Since I am

[2] We shall examine non-anachronistic examples in Chapter 8.

arguing that time ought to be understood as a compound of motion and perception, the question concerning the perception of the common perceptibles naturally arises, and its answer is both surprising and revealing. In the opening chapter of *De Anima* III, Aristotle claims that we perceive *all* of the common perceptibles by movement (425a15 ff.). I endeavor to determine exactly what Aristotle has in mind in making this claim, which requires examining the role that *phantasia* plays within anticipation and memory.

As we have seen, Aristotle states his official definition of time in *Physics* IV.11 at 219b1: "For that is what time is: a number of motion with respect to the before and after" (*touto gar estin ho chronos, arithmos kinêseôs kata to proteron kai husteron*).[3] This formula is a choice example of Aristotle's characteristically compact definitions in that it highlights the central elements of a fully developed theory, but which is not terribly informative in isolation. Obscurity of the definition aside, it is clear from the nature of the discussion in *Physics* IV.11–14 that Aristotle is offering a real definition, not a nominal definition. That is, he's giving us what he takes to be the essential nature of time itself rather than an account of how the word "time" (rather, "*chronos*") is conventionally used.[4]

Aristotle places time in the genus of *number*. Specifically, it is a number of motion. The final qualification links it with the before and after. I argued in the preceding chapter that the phrase "with respect to the before and after" in the definiens refers to kinetic cuts. The argument, in brief, was this: on pain of circularity, "before" and "after" cannot be used there in any temporal sense; and given the fact that motion is mentioned in the definiens, the most obvious alternative candidate is the before and after in motion. The question we must now face, then, is what Aristotle intends when he claims that time is a number associated with kinetic cuts.

Independently of any theoretical motivation, Aristotle had good reason to think of time in terms of numbers and movements. Recall from Chapter 2 that the prevailing convention for dating years in Aristotle's time employed the form of "the Nth year relative to event *E*," where the event in question was more or less determined by contextual salience. Thus we have Thucydides' use of "in the summer of the third year of this war" and

[3] τοῦτο γάρ ἐστιν ὁ χρόνος, ἀριθμὸς κινήσεως κατὰ τὸ πρότερον καὶ ὕστερον. Ross notes that MS J and Simplicius omit "τὸ," while all others retain it. I concur with Ross' retention of the article and suspect that Simplicius' omission of it is related to his failure to properly understand the nature of kinetic cuts (see section 5.1.1).

[4] Compare *An. Post.* II.10 with *Top.* 153a15 ff. and *Met.* 1031a1 ff. for contrasting discussions of nominal definitions and real definitions, respectively.

Herodotus' "in the eleventh year of their exile." There was, as we noted, the conventional use of the four-year cycle of the Panhellenic games as a standard period (the "Olympiad"), and the treatment of the first Olympic games in 776 BCE as an epoch. Thus, "the thirtieth Olympiad" would designate a period spanning the years 656–653 BCE. Such dating practices would have provided just the sort of common wisdom Aristotle is fond of treating as a foundation for his philosophical views and certainly had some sway in his conception of time as a number of motion.

However, although he is anxious to preserve the judgments of common wisdom, he invariably builds a sophisticated theoretical apparatus on top of it. So we cannot expect that this is anything more than the beginning of the complete story. The question, again, is how to understand the relevant sense of "number" as it relates to kinetic cuts, the before and after in motion.

The passages that immediately precede the official definition bristle with occurrences of "before" and "after," so we would do well to consider the various ways in which they are being used there. The passage beginning at 219a14 lays out the tripartite division of the varieties of before-and-after that were discussed in Chapter 5: in spatial magnitude, in motion, and in time. The passage that runs from 219a22 up to the official definition is worth examining in its entirety:

> Additionally, we also come to know time whenever we mark off [*horisômen*] motion, marking it off by means of [i] the before and after; and we say that time has passed whenever we have a perception [*aisthêsin*] of [ii] the before and after in motion. We mark out motion by taking [*hupolabein*] [iii] them to be different from one another, and different from another thing between them. For whenever we conceive [*noêsômen*] of the limits as different from the middle, and the soul says that there are two nows, the one [iv] before and the other after, that's when and that's what we say time is; for what's marked off by the now is thought to be time – let this be granted. So whenever we perceive the now as one, and not as [v] before or after in the motion, or as the same but pertaining to something which is [vi] before and after, no time seems to have passed, because no motion seems to have taken place, either. But whenever we do perceive [vii] the before and after, that's when we speak of time.[5]

[5] ἀλλὰ μὴν καὶ τὸν χρόνον γε γνωρίζομεν ὅταν ὁρίσωμεν τὴν κίνησιν, τῷ πρότερον καὶ ὕστερον ὁρίζοντες· καὶ τότε φαμὲν γεγονέναι χρόνον, ὅταν τοῦ προτέρου καὶ ὑστέρου ἐν τῇ κινήσει αἴσθησιν λάβωμεν. ὁρίζομεν δὲ τῷ ἄλλο καὶ ἄλλο ὑπολαβεῖν αὐτά, καὶ μεταξύ τι αὐτῶν ἕτερον· ὅταν γὰρ ἕτερα τὰ ἄκρα τοῦ μέσου νοήσωμεν, καὶ δύο εἴπῃ ἡ ψυχὴ τὰ ηῦν, τὸ μὲν πρότερον τὸ δ' ὕστερον, τότε καὶ τοῦτό φαμεν εἶναι χρόνον· τὸ γὰρ ὁριζόμενον τῷ νῦν χρόνος εἶναι δοκει· καὶ ὑποκείσθω. ὅταν μὲν οὖν ὡς ἓν τὸ νῦν αἰσθανώμεθα, καὶ μὴ ἤτοι ὡς πρότερον καὶ ὑστέρον ἐν τῇ κινήσει ἢ ὡς τὸ αὐτὸ μὲν προτέρου δὲ καὶ ὑστέρου τινός.

As I have numbered them, there are seven occurrences of "the before and after" (or a cognate expression) in this passage. The intended sense of several occurrences poses little difficulty. Occurrence [ii] is an explicit reference to kinetic cuts, and this fact strongly suggests that occurrence [i] functions similarly. The anaphoric occurrence in [iii] can only refer back to occurrence [ii], so this must also be a reference to kinetic cuts. Occurrences [iv] and [v] are adjectival, modifying "the now(s)" in each case, for which reason we should probably take these to be temporal and to refer to instants. The remaining occurrences are more challenging to interpret. Occurrence [vi] is adjectival, but the antecedent of the pronoun being modified is not obvious. The final occurrence is situated within the context of perceiving the before and after. Unfortunately, both kinetic cuts and instants appear as perceived items in the passage, so neither can be ruled out decisively. Since the two previous uses of "before and after" seem to be references to instants, there's some reason for thinking that this last occurrence also functions temporally.

On balance, the passage demonstrates just how comfortable Aristotle feels shifting between the kinetic and temporal senses of the expression. I shall provide an explanation for this attitude below, but more to the immediate point is Aristotle's abundant use of cognitive language. He speaks of marking out, perceiving, taking, and conceiving. Recall, too, that the opening argument of IV.11 (which I dubbed the "Exactly When" argument) is couched similarly in terms of noticing, perceiving, and marking off. So given the abundant use of such cognitive language in the vicinity, why does Aristotle define time as a kind of *number*?

6.1 "*ARITHMOS*" IS SAID IN MANY WAYS

Prima facie, the expression "a number of motion" denotes some sort of quantity. In *Metaphysics* Δ.13, Aristotle defines *quantity* (*poson*) as "that which is divisible into two or more constituent parts of which each is by nature a one and a this" (1020a7–8).[6] *Quantity* is a genus that divides into two species: *plurality* (*plêthos*), and *magnitude* (*megethos*). Aristotle gives two sets of differentiae: first, pluralities are numerable (*arithmêton*), while magnitudes are measurable (*metrêton*) (a8–10); second, pluralities

οὐ δοκεῖ χρόνος γεγονέναι οὐδείς, ὅτι οὐδὲ κίνησις. ὅταν δὲ τὸ πρότερον καὶ ὕστερον, τότε λέγομεν χρόνον·

[6] Ποσὸν λέγεται τὸ διαιρετὸν εἰς ἐνυπάρχοντα ὧν ἑκάτερον ἢ ἕκαστον ἕν τι καὶ τόδε τι πέφυκεν εἶναι.

are divisible into non-continuous parts, while magnitudes are divisible into continuous parts (a10–11).[7]

Let us consider a toy example that illustrates the intended distinction. Imagine a circular horse corral whose diameter is three decameters. The length of its diameter is divisible into thirty meters, each of which is divisible into ten decimeters, which are each divisible into ten centimeters, and so on. The continuity of the spatial interval guarantees that as we divide, we never reach a stage at which the results of division are not themselves further divisible into yet further divisible items (which may or may not have a special name) that belong to the same kind as that from which they originated – namely, spatial intervals. The diameter of the corral is a magnitude: it is a continuous whole that we quantify by measuring, i.e. by giving the number of multiples of some conventionally defined sub-segment whose fusion exhausts the whole. There is no deep metaphysical difference between a meter and a centimeter; the former is nothing more than the fusion of one-hundred multiples of the latter. A consequence of this is that there is no natural unit of spatial magnitude; we are free to select any arbitrary quantity that suits our needs and treat it as a standard.

Now suppose that eight horses inhabit the corral. The group of horses is divisible into two groups of four, each of which is divisible into two duos, which are each further divisible into two singleton "groups." Once we have reached this level, we can divide no further without producing items of a different kind – namely, *horse parts.* A group of horses is more than just so much flesh, blood, bone, and hair; it is something that divides evenly into natural units, in the fully Aristotelian sense of "natural". That is to say, dividing the group of eight into four or twelve "units" results in items whose metaphysically derivative status is reflected in the language we use to characterize them: a *pair of* horses, *two-thirds of* a horse, and so on. A group of horses is a plurality: it is a collection of wholes that we quantify by numbering – i.e. by individuating the members (in this case, natural-kind members) that constitute the group

7 Cf. Ross (1936), 69: "A thing is continuous [*sunechēs*] when the converging extremes of its parts not merely are ἅμα [together] but are ἕν [one]." In general, then, Aristotle's concept of continuity is that of an undivided but divisible whole which, were it to be divided, would produce two undivided but divisible tokens of the original type. A slightly different way to put the point is to say that no two divisions are contiguous: between any two divisions there lies a type-segment which is itself infinitely divisible. As many have pointed out before, this characterization is not precise enough to distinguish the set of real numbers from the set of rationals, but for Aristotle's purposes, density is sufficient. For a detailed discussion of Aristotle's conception of continuity see Sorabji (1983), esp. ch. 21; Corish (1969); and Bostock (1991).

when "corralled together," setting them in a one-to-one relation with the natural numbers.

While this distinction is made evident enough by such toy examples, it raises an apparent difficulty for the view Aristotle develops in *Physics* IV, for he maintains in connection with time that both measurability and numerability are involved. Consider the following passage:

And so time isn't motion, but is motion insofar as it has a number [*arithmon*]. A sign of this: we determine the greater and smaller by means of number, and with respect to motion, greater and smaller is determined by means of time. Hence, time is a certain kind of number. But since "number" is ambiguous (for we speak of number both as the thing numbered and the numerable, as well as that with which we number), time is in fact the thing numbered and not that with which we number. (219b2–8)[8]

One function of this passage is to discourage us from thinking of time as an abstract number, as something that might exist independently of individual movements and moving individuals. Aristotle's staunch anti-Platonism with respect to the metaphysical status of numbers (a position developed in detail in *Metaphysics* I.1–3)[9] does considerable work in keeping time tethered to the perceptible world. More relevant to the present issue is the fact that, because Aristotle claims that time numbers motion, time seems to belong to the species *plurality*, the kind of quantity that is numerable (*arithmêton*).

There are, however, several points at which Aristotle connects motion and time by means of a measuring relation, which should incline us to think that time properly belongs in the species *magnitude*:

Not only do we measure (*metroumen*) the movement by the time, but also the time by the movement. (220b14–16)[10]

Time is a measure (*metron*) of motion and of being moved. (220b32–3)[11]

But since time is a measure (*metron*) of motion, it will also be a measure of rest. (221b7–8)[12]

So what is Aristotle's view? Is time a magnitude or a plurality? The answer seems to be *both*, though equivocally:

[8] οὐκ ἄρα κίνησις ὁ χρόνος ἀλλ' ᾗ ἀριθμὸν ἔχει ἡ κίνησις. σημεῖον δέ· τὸ μὲν γὰρ πλεῖον καὶ ἔλαττον κρίνομεν ἀριθμῷ, κίνησιν δὲ πλείω καὶ ἐλάττω χρόνῳ· ἀριθμὸς ἄρα τις ὁ χρόνος. ἐπεὶ δ' ἀριθμός ἐστι διχῶς (καὶ γὰρ τὸ ἀριθμούμενον καὶ τὸ ἀριθμητὸν ἀριθμὸν λέγομεν, καὶ ᾧ ἀριθμοῦμεν), ὁ δὴ χρόνος ἐστὶν τὸ ἀριθμούμενον καὶ οὐχ ᾧ ἀριθμοῦμεν.

[9] See Annas (1975) for a nice treatment of how this distinction – and the distinction between numbering and measuring, such as it is – plays out in Aristotle's discussion of time in *Phys.* IV.

[10] οὐ μόνον δὲ τὴν κίνησιν τῷ χρόνῳ μετροῦμεν, ἀλλὰ καὶ τῇ κινήσει τὸν χρόνον.

[11] ὁ χρόνος μέτρον κινήσεως καὶ τοῦ κινεῖσθαι.

[12] ἐπεὶ δ' ἐστὶν ὁ χρόνος μέτρον κινήσεως, ἔσται καὶ ἠρεμίας μέτρον.

> The smallest number without qualification is two; but in a particular case, there is in one sense a smallest number, but in another sense not. For instance, of a line, the smallest in plurality is two lines, or one line, but there is no smallest in magnitude. For every line can be everywhere divided. And so time is also like this. With respect to number, the smallest is one time, or two times, but there is no smallest time with respect to magnitude. (220a27–32)[13]

The implicit contrast here is between the notions of *a time* and *time* – or to put the point slightly differently, between *units* of temporal extension on the one hand, and temporal extension *in general* on the other.[14] Units of temporal extension are numbered or counted. Considered as a series or collection of discrete temporal intervals (days, months, seasons), time is a plurality. In contrast, temporal extension considered generally is a continuous magnitude that is measured. It (or some finitely long segment of it) is measured in the same way that the diameters of circular corrals are measured: by giving the number of multiples of an arbitrary, but commonly recognized, sub-segment-type whose fusion exhausts the whole. This distinction deserves detailed examination, so let us approach the issue by considering more carefully the significance of time as a kind of number specifically as it bears on the hylomorphic interpretation.

6.2 PERCEIVING, DIVISION, AND DETERMINACY

One way in which Aristotle characterizes matter explicitly invokes the notions of number and quantity. His mature hylomorphism is developed in *Metaphysics* Z and H, and in Z.3 he tells us that "what I mean by matter per se is something that is neither a this, nor of a particular quantity" (1029a20–1).[15] In this context he is thinking principally of elemental matter and the higher-level combinatorial products thereof. But given the iterative nature of the matter–form distinction, the point is quite general: for any kind of individuable composite *C*, the material component of any specimen *C* has no determinate status with respect to the predicate *C*. Or, put more simply, to be *C*-type matter is to be *determinable*

[13] Ἐλάχιστος δὲ ἀριθμὸς ὁ μὲν ἁπλῶς ἐστὶν ἡ δυάς· τὶς δὲ ἀριθμὸς ἔστι μὲν ὡς ἔστιν, ἔστι δ' ὡς οὐκ ἔστιν, οἷον γραμμῆς ἐλάχιστος πλήθει μέν ἐστιν αἱ δύο ἢ ἡ μία, μεγέθει δ' οὐκ ἔστιν ἐλάχιστος·

[14] Cf. White (1989). While I find White's discussion of Aristotle's analysis of the identity conditions for *a time* to be generally on the mark, he draws the distinction too deeply to fit Aristotle's view – White says that he's clarifying a distinction that Aristotle isn't sufficiently sensitive to (208) – and ultimately spoils his interpretation of Aristotle's account of *time*.

[15] λέγω δ' ὕλη ἣ καθ' αὑτὴν μήτε ετὶ μήτε ποσὸν.

by means of the form *C*: form is what *makes* a certain heap of matter an individual of the relevant type.

I want to suggest that Aristotle uses "number" in his definition of time to implicate just this sort of determinacy. A number of motion, in this sense, is a determinate, unitary movement. Movements are essentially continuous magnitudes. Any particular movement is one in the sense that to be a movement is to be the actuality of a particular *kinoumenon*, upon whose unity the unity of the movement is grounded.[16] Insofar as the existence of a *kinoumenon* is fully objective and determinate, each movement that actually occurs is itself objectively fully determinate.[17]

But motion, as a ubiquitous natural phenomenon, is not so determinate. Motion has always been taking place, on Aristotle's view, so there is no beginning or end to serve as defining limits of a determinate unity.[18] Furthermore, as a continuous magnitude, any particular movement is such that it is infinitely divisible; it is a peculiar unity-in-plurality. As Aristotle remarks in *Physics* VI.6, "since the nows are infinite [*apeira*] in number, everything that is moving must have completed an infinite number of movements" (237a15–17).[19] He qualifies this position in VIII.8, however, by maintaining that the infinitely many sub-movements are merely potential rather than actual:

> So in the straight line any one of the points lying between the two extremes is potentially a middle-point; but it is not actually so unless that which is in motion divides the line by coming to a stand at that point and beginning its motion again; thus the middle-point becomes both a beginning and an end, a beginning of the latter part and an end of the first part. (262a22–6)[20]

This view comprises part of Aristotle's solution to Zeno's paradoxes of motion. The core idea is that an actual movement is a well-defined bit of motion. Any of the points along a moving object's trajectory *could* serve as a defining limit of a sub-movement, but in order for it to actually serve as such, the moving object must pause at the point of division. But in fact, coming to rest doesn't divide the larger movement – it *completes* the smaller movement and sets up the necessary conditions for the initiation

[16] Recall *Phys.* IV.11, 220a6–8.

[17] Each movement is *individually* determinate, but the relations obtaining between movements (e.g. simultaneity) are not objectively determinate in the mind-independent sense of "objective." I take up this issue in Chapter 11.

[18] We shall examine the textual basis for this claim in detail in Chapter 11.

[19] ὥστ' εἰ τὰ νῦν ἄπειρα, πᾶν τὸ μεταβαλλον ἄπειρα ἔσται μεταβεβληκός.

[20] ὥστε τῆς εὐθείας τῶν ἐντὸς τῶν ἄκρων ὁτιοῦν σημεῖον δυνάμει μέν ἐστι μέσον, ἐνεργείᾳ δ' οὐκ ἔστιν, ἐὰν μὴ διέλῃ ταύτῃ καὶ ἐπιστὰν πάλιν ἄρξηται κινεῖσθαι· οὕτω δὲ τὸ μέσον ἀρχὴ γίγνεται καὶ τελευτή, ἀρχὴ μὲν τῆς ὕστερον, τελευτὴ δὲ τῆς πρώτης.

of a *distinct* movement. How, then, is the notion of division to be understood in this context?

The answer is given very shortly after the passage quoted above, where Aristotle explains why coming to a stand makes the smaller movement actual:

> When *B*, the middle-point, serves both as an end and as a beginning for the *pheromenon A*, *A* must come to a stand at *B*, because doing so makes it two just as one might do in thought. (262b5–6)[21]

Aristotle elucidates this idea of making a point two in thought less than one page later:

> In dividing the continuous distance into two halves, one point is treated as two, since we make it a beginning and an end. And this same result is produced by the act of counting halves as well as by the act of dividing into halves. (263a23–6)[22]

To divide a continuous movement, then, an agent treats a single point as two: it is at once the end of the first sub-movement and the beginning of the second sub-movement. Now, since the sort of division we're investigating is division of a movement, we ought to try to understand this idea in kinetic terms – specifically, within the metaphysical structure of kinetic cuts. As a first attempt, consider the following pair of descriptions:

<!*kin*(*C*), *B*>
<!*kin*(*C*), *B**>

These differ from one another by treating the point of division, *B*, as two. Although this strategy hews closely to Aristotle's words, it pretty obviously fails. Because the two descriptions share a common *kinoumenon* as their first element, it follows definitionally that there is a single movement that includes points *B* and *B**. Treating the point as two is no more effective in creating a division within the movement than is recording the fact that one and the same *kinoumenon* occupies many diverse points within a movement. So this strategy is inadequate.

The following might seem more promising:

<!*kin*(*C*), *B*>
<!*kin*(*C*)*, *B*>

21 ὅταν δὴ χρήσηται τὸ φερόμενον Α τῷ Β μέσῳ καὶ τελευτῇ καὶ ἀρχῇ, ἀνάγκη στῆναι διὰ τὸ δύο ποιεῖν, ὥσπερ ἂν εἰ καὶ νοήσειεν.

22 ἐὰν γάρ τις τὴν συνεχῆ διαιρῇ εἰς δύο ἡμίση, οὗτος τῷ ἑνὶ σημείῳ ὡς δυσὶ χρῆται· ποιεῖ γὰρ αὐτὸ ἀρχὴν καὶ τελευτήν. οὕτω δὲ ποιεῖ ὅ τε ἀριθμῶν καὶ ὁ εἰς τὰ ἡμίση διαιρῶν.

Here we have conceptually distinct *kinoumena*, which is certainly sufficient to distinguish the two sub-movements from one another. But this approach suffers the opposite defect from the one examined above: whereas it hewed closely to Aristotle's words but was inadequate, this one is adequate but strays too far from Aristotle's words. Which "point" here is being treated as two? Evidently, none, since the *kinoumenon* is fulfilling dual roles.[23]

I believe that Aristotle did not express his intention as clearly as he ought to have in the passage in question. But his intention is clear enough: just as points divide lines and instants divide temporal intervals, movements are divided by kinetic entities – that is, by kinetic cuts:

<!*kin*(*C*), *B*>
<!*kin*(*C*), *B*>*

In dividing a movement by treating one point as two, the agent treats the moving-thing-at-*B* as two. Recall the passage at 220a18–20 that was examined in Chapter 5: "Further, it is also clear that the now is no part of time, nor is the division [*diairesis*] a part of the movement, just as the point is not part of the line."[24] The "division" in question is a kinetic cut, the kinetic analogue of a point. Conceptual division of a movement is effected by treating one of these "kinetic points" as two.

So a movement contains infinitely many (potential) sub-movements insofar as any kinetic cut may be treated as the end of the prior movement and the point-as-from-which the posterior movement begins.[25] This account bears a strong affinity with one of Aristotle's remarks in *Physics* IV.

[T]he now is the before and after [in motion] *qua* numerable. (219b25, repeated at b28)[26]

I believe that the sense of "numerable" is now clear: its meaning is roughly that of "individuable," and the individuation in question is effected by perception or thought. This reading of "numerable" is strongly

[23] I set aside the strategy according to which the *telos* is treated differently: <!*kin*(*C*), *B*>, <!*kin*(*C**), *B*>. Although this approach treats a point as being two, it treats the wrong point (the *telos* rather than the point of division) as two.

[24] καὶ ἔτι φανερὸν ὅτι οὐδὲν μύριον τὸ νῦν τοῦ χρόνου, οὐδ᾽ ἡ διαίρεσις τῆς κινήσεως, ὥσπερ οὐδ᾽ ἡ στιγμὴ τῆς γραμμῆς.

[25] Recall that there is no first kinetic cut in a movement (see section 5.4).

[26] ᾗ δ᾽ ἀριθμητὸν τὸ πρότερον καὶ ὕστερον, τὸ νῦν ἔστιν. Aristotle slips at 223a28–9 when he says, "The before and after are within motion, and time is these *qua* numerable" (τὸ δὲ πρότερον καὶ ὕστερον ἐν κινήσει ἐστίν· χρόνος δὲ ταοῦτ᾽ ἐστὶν ᾗ ἀριθμητά ἐστιν). Given his earlier remarks in *Phys.* IV.11, he ought rather to have said "time is what is bounded by these *qua* numerable."

supported by the passage immediately preceding Aristotle's definition of time (examined in the opening section of this chapter) that employs the notions of marking off motion, perceiving the before and after in motion, and conceiving of limits as different from what lies between them (219a22–9). It would be perverse to take the notion of numerability to be about counting in any explicit way, for Aristotle surely isn't telling us that we perceive time only when we engage in acts of subvocalized counting.[27]

But now we must address a troubling ambiguity in Aristotle's view. He defines time as *a number* of motion, and yet his characterization of the now is always couched in the weaker terms of *numerability*. Why the difference?

We have already seen this difference manifest in the distinctions between counting and measuring, on the one hand, and between determinacy and determinability, on the other. Every movement is in time and has a time, even if no one happens to perceive it, mark it off, or otherwise conceive of it.[28] Similarly, every kinetic cut is a now, even if it goes unnoticed by any percipient agent. So when Aristotle says that time is a number of motion, he is saying (minimally) that time is motion's determinability, its susceptibility to being divided into determinate quantities, undetached parts of arbitrary length. This is a property that motion would not have but for the capacities and activities of percipient souls.[29]

On the other hand, *a time* is a determinate quantity of motion, an actual division of some movement that is effected by explicitly marking out kinetic cuts as defining limits. Recall the passage at 220a27–33 quoted in the preceding section:

[27] Cf. Ross (1936, 65), but *contra* Coope (2005, 91). Also cf. *Met.* I.1: "Knowledge also, and perception, we call the measure of things, for the same reason, because we know something by them" (καὶ τὴν ἐπιστήμην δὲ μέτρον τῶν πραγμάτων λέγομεν καὶ τὴν αἴσθησιν διὰ τὸ αὐτό, ὅτι γνωρίζομέν τι αὐταῖς) (1053a31–2).

[28] See, e.g., 222b30–1: "Now that we've drawn these distinctions, it's clear that every change and every *kinoumenon* is in time" (Τούτων δ' ἡμῖν οὕτω διωρισμένων φανερὸν ὅτι πᾶσα μεταβολὴ καὶ ἅπαν τὸ κινούμενον ἐν χρόνῳ). The notions of being in time and having a time will be discussed in Chapter 11 within the context of an analysis of simultaneity.

[29] Whence Aristotle's remark in *Phys.* IV.14 that time would not exist if there were no counting souls. Cf. *Met.* Δ.15: "Relative terms which imply number or capacity, therefore, are all relative because their very essence includes in its nature a reference to something else, not because something else is related to it, but that which is measurable or knowable or thinkable is called relative because something else is related to it. For the thinkable implies that there is a thought of it" (τὰ μὲν οὖν κατ' ἀριθμὸν καὶ δύναμιν λεγόμενα πρός τι πάντα ἐστὶ πρός τι τῷ ὅπερ ἐστὶν ἄλλου λέγεσθαι αὐτὸ ὅ ἐστιν, ἀλλὰ μὴ τῷ ἄλλο πρὸς ἐκεῖνο· τὸ δὲ μετρητὸν καὶ τὸ ἐπιστητὸν καὶ τὸ διανοητὸν τῷ ἄλλο πρὸς αὐτὸ λέγεσθαι πρός τι λέγονται. τό τε γὰρ διανοητὸν σημαίνει ὅτι ἔστιν αὐτοῦ διάνοια) (1021a26–31). See Hintikka *et al.* (1977).

The smallest number without qualification is two; but in a particular case, there is in one sense a smallest number, but in another sense not. For instance, of a line, the smallest in plurality is two lines, or one line, but there is no smallest in magnitude. For every line can be everywhere divided. And so time is also like this. With respect to number, the smallest is one time, or two times, but there is no smallest time with respect to magnitude.[30]

Putting the pieces together, we now see that *time* (determinable temporal extension) is or can be measured, whereas *times* (determinately long undetached parts of temporal extension) are counted. In both cases, motion serves as the matter of time by virtue of its intrinsic indeterminacy with respect to length, and perception serves as the form of time by (i) (always) contributing one of the metaphysically necessary conditions for motion's determinability, and (ii) (sometimes) actively imposing determinacy upon individual movements.

The hylomorphic conception of time as motion enformed by perception explains Aristotle's abundant use of cognitive language in the discussion that leads up to the official definition at 219b1. The passage at 219a22 cited above is especially illustrative of this. Moreover, this interpretation also explains the ease with which Aristotle alternates between the kinetic and temporal senses of "before" and "after" in this passage. Unless one is discussing kinetic cuts in very abstract terms (as I did, for example, in the preceding chapter), one cannot help but conceive of them as instants in time. In fact, this is Aristotle's explanation for what we noted in the opening section of this chapter to be a feature of our psychological makeup. When someone imagines a particular movement, he cannot fail to imagine what Aristotle calls "the before and after in motion," or kinetic cuts. But according to Aristotle's account, as they are perceived (or conceived or imagined) by a sentient being, kinetic cuts just are instants. This is why we find it difficult, though not impossible, to conceive of motion atemporally.

In addition to the preceding, this interpretation helps us to make sense of other elements of Aristotle's account that would otherwise be quite puzzling. One of these is the inference at the opening of *Physics* IV.11 that we examined in Chapter 3. Aristotle argues that, since we notice time when and only when we notice motion, time must be some aspect of motion. We saw that this inference goes through only if he supposes prior to giving any argument that time is an evident proper feature of motion. We

[30] As an aside, Aristotle says "or two times" as a concession to the view, held by some ancients, that *one* is not a number, properly speaking, since (they believed) "number" denotes plurality.

may now examine how this aspect of his view relates to the distinction between determinability and determinacy.

Recall the nature of the evident proper feature relation:

(EPF) For any pair of type-perceptibles <Φ, Ψ>, Φ is an *evident proper feature* of Ψ iff every token Ψ_i (and nothing else, except by virtue of its relation to some such token) has some token Φ_i as one of its features, and for any percipient *z*, *z* cannot perceive any Ψ_i *as* a Ψ without also perceiving its Φ_i as a Φ.

Construed as the determinability of motion, time clearly is an evident proper feature of motion. To perceive a movement as such is to perceive it as containing indefinitely many kinetic cuts. A kinetic cut, insofar as it is perceptible, just is a now. But "whenever … the soul says that there are two nows … that's when and that's what we say time is" (219a26–9).[31] So perceiving a movement as such (a suitable instance of "Ψ_i *as* a Ψ") invariably requires perceiving it as having (being in or associated with) some particular time (its Φ_i).

Here an analogy with spatial perception might prove to be helpful. A person can "measure" a natural scene in which he finds himself simply by scanning the landscape between himself and the horizon (so called for reasons directly relevant to the present discussion). He hasn't done what a surveyor would do with her transit and tape measure, but he has "counted" the vista in the sense that he has taken account of it. Taking one's standing position and the horizon to define a spatial interval is a way of measuring it in the sense of countenancing the interval's determinability with respect to length. Indeed, the landscape need not even be real: if one perceives a painted landscape *as* a landscape, one must assign (in a way to be discussed more fully in Chapter 9) some value of distance to the painted horizon. Analogously, simply noticing two distinct kinetic cuts within a particular movement is sufficient to measure the movement in this thin sense. This explains why Aristotle believes that time is an evident perceptible feature of motion: the very act of perceiving a movement *as* a movement requires that one perceive the movement as having some indeterminate (but determinable) value of extension. And on Aristotle's view, this value of extension for a movement, insofar as it is perceptible, is time. To be aware of time, in this sense, is simply to take account of some definite period of motion, which Aristotle himself notes at the opening of IV.11 is something that a person can't help but do, provided that she isn't unconscious.[32]

[31] ὅταν … δύο εἴπῃ ἡ ψυχὴ τὰ ηῦν … τότε καὶ τοῦτό φαμεν εἶναι χρόνον·
[32] More on this in Chapter 7.

In addition to the thin sense of measuring associated with determinability, there is also a thicker sense associated with determinacy. This thicker sense of measuring motion consists of three central elements: (i) precisely defining movements (or sub-movements), (ii) the comparison of two (sub-) movements with one another, and (iii) the treatment of one of the (sub-) movements as a standard.[33] In point of fact, a particular movement is treated as a standard only to the extent that it is regarded as a token of some suitable type of motion.[34] Once a motion type is established as a standard, one measures other movements by counting the number of standard movements corresponding to the other.[35] Let us examine how this notion fits into the (EPF) relation.

In his description of court proceedings in the *Constitution of Athens*, Aristotle remarks that:

> Ten *choes* [of water] are allowed for a case in which an amount of more than five thousand drachmas is involved, and three for the second speech on each side. (67.2)

The *chous* is itself a standard of liquid measurement, so its adoption as a standard for measuring time is principally a matter of convenience. Setting aside such practical considerations, we may say that the relevant type Ψ in question is the voiding of one *chous* by a standard-design *klepshydra*. Any particular voiding of one *chous* is the relevant token of Ψ, a concrete movement Ψ_i. Call the motion type Ψ "*C–V*", and call any token of the type a "*chous* voiding." Now since *C–V* is a *standard* motion type, every *chous* voiding has its kinetic length essentially (since that's just what it is to be a token of a standard type).[36] So to perceive a *chous* voiding as an instance of *C–V* (Ψ_i *as* a Ψ) is to perceive a particular lapse of a standard

[33] David Bolotin (1997) offers a similar treatment of the topics of perception and measurement. However, Bolotin stumbles on the question of the appropriate size of kinetic positions and the now. He characterizes kinetic positions as "coincidences between stages of … different motions" (52), and he takes the now of IV.10–14 to be an interval (60–1), in spite of the claim – which Bolotin acknowledges – in *Phys.* VI.3 that the now properly so called is indivisible. The implication of the latter point for Bolotin's interpretation is that Aristotle uses "now" in some derivative sense even while framing his theory of time, which is implausible.

[34] The current definition of "the second" illustrates this nicely. In 1967 the Thirteenth General Conference on Weights and Measures defined the second of atomic time in the International System of Units (SI) as "the duration of 9,192,631,770 periods of the radiation corresponding to the transition between the two hyperfine levels of the grade state of the cesium-133 atom."

[35] The correspondence relation in question here is obviously simultaneity. I shall provide a detailed analysis of simultaneity in Chapter 11.

[36] It would be desirable if Aristotle had available the theoretical resources to secure the existence of Ψ as a genuine type, and thereby to guarantee the equivalence of distinct Ψ_i. I shall show in Chapter 11 that he does.

period of time Φ_i. Certainly, one may perceive a *chous* voiding merely as the emptying of a container of water (as a child or someone untutored in court proceedings might), but anyone who perceives the event as a token of the relevant type will thereby perceive it as a timekeeping event.

The clarity and plausibility of this sort of example lends significant support to my reconstruction of the "No Motion without Time" argument in Chapter 3, which makes use of the (EPF) relation. What remains to be seen (in Part IV) is how acts of perception manage to precisely mark out movements like *chous* voidings.

Before moving on to consider an extended challenge to this interpretation, I want to elaborate on the analogy between space and time I employed in our examination of the thin (determinability) sense of measuring motion.

In fact, I don't intend the comparison to be a mere analogy; as I understand Aristotle, distance is to space *precisely* as time is to motion. Both space and motion are continuous magnitudes. Because there are percipient agents equipped with faculties suited to perceiving space and motion, each of them is susceptible to being "carved up" in various ways by those agents. The use of "carving" here is metaphorical, of course, since neither space nor motion can be rendered asunder by human activities. The divisions are conceptual, and the parts thereby defined are undetached from the whole. A remark Aristotle makes at 220b32–222a4 confirms these comparisons:

> [A] time is a measure of moving and of being moved, and it measures out the movement by means of defining some movement which can measure the whole (just as the cubit measures the length by defining some magnitude which can measure the whole).[37]

Despite the exact analogy in the relevant respects, the comparison requires us to acknowledge a difference between measuring spatial magnitudes and measuring movements: the former can be measured along three dimensions, whereas the latter can be measured only along one dimension. We can sensibly ask how long a movement is (or how long it takes), but not how wide or how high it is. Aristotle has a very good explanation as to why time exists only along a single dimension, one that turns on the structure of motion and change. Since every predicate has just one maximally different contrary, there is but one dimension along which any

[37] ἐπεὶ δ' ἐστὶν ὁ χρόνος μέτρον κινήσεως καὶ τοῦ κινεῖσθαι, μετρεῖ δ' οὗτος τὴν κίνησιν τῷ ὁρίσαι τινὰ κίνησιν ἣ καταμετρήσει τὴν ὅλην (ὥσπερ καὶ τὸ μῆκος ὁ πῆχυς τῷ ὁρίσαι τι μέγεθος ὃ ἀναμετρήσει τὸ ὅλον). Notice that I have supplied the indefinite article.

particular change can take place. Something green cannot become something sweet except coincidentally, in the way that a green apple that happens to be tart becomes a sweet apple that happens to be red.[38] So change takes place only along one axis: to be a *kinoumenon* is to be an *F*-type *kinoumenon*, for some particular predicate *F*.[39] This is why it makes no sense to ask how wide or how high a movement is, even though it does make sense to ask how long it is.

Let us turn now to a potent challenge to the line of interpretation that I have been developing.

[38] Cf. *Phys.* 1.5, 188b4 ff.

[39] While we nowadays analyze curvilinear motion as a compound of two (or three, in some cases) simple rectilinear movements, each of which takes place along a single spatial axis, Aristotle treats rectilinear and rotational motion as fundamentally different types. This is significant for his view of time, as we'll see in Chapter 11.

CHAPTER 7

On a moment's notice

It should be clear that in order for the hylomorphic interpretation of Aristotle's temporal theory to be viable, it is imperative that Aristotle have the theoretical resources to account for the perception of kinetic cuts. For without such, he will not be able to maintain (as he explicitly does in *Physics* IV.11) that we actually do perceive the now as before and after in motion.

To address this challenge, we must venture beyond the pages of the *Physics*. The text bearing most directly on this issue is chapter 7 of the *De Sensu*, in which Aristotle declares with confidence: "[N]or can there be any imperceptible time, none that escapes our notice; rather, it is possible to perceive each [*pantos*]" (448a25–7).[1] The antecedent of "each" is a matter of some controversy. The most proximate candidate referent in the passage is "time" (*chronos*), but several translations read "instant" (*nun*) into the text.[2] Neither can be ruled out on grammatical grounds. If we accept the former reading and bear in mind Aristotle's stated position in *Physics* IV.11 that "the now is no part of time" (220a18–19),[3] then the remark seems in no way to commit Aristotle to the perceptibility of instants.

Given the naturalness of specifying "time" as the antecedent of "each," why do many translators prefer "instant"? One possible reason might be the desire to find somewhere in the text an independent argument for the perceptibility of instants, for that kind of argument would further fortify the theory of time developed in the *Physics*. However, I rather doubt that this is the motivating reason for those translators who prefer "instant"

[1] οὐδ' ἐνδέχεται χρόνον εἶναι ἀναίσθητον οὐδένα οὐδὲ λανθάνειν, ἀλλὰ παντὸς ἐνδέχεται αἰσθάνεσθαι.

[2] So Beare in Barnes (1984); Hett (1936); and Ross (1973). Thanks to Richard Bett for making the salience of this issue clearer to me in his comments on a paper I delivered at the 2007 Pacific Division meetings of the American Philosophical Association, on which this section is largely based.

[3] καὶ ἔτι φανερὸν ὅτι οὐδὲν μόριον τὸ νῦν τοῦ χρόνου.

to "time," because this portion of the *De Sensu* has, to the best of my knowledge, never been invoked in support of Aristotle's temporal theory.[4] This is a surprising fact, given the apparent connection between the two lines of inquiry. But as I have indicated at several points in this book, the role of perception in Aristotle's account of time simply has not been adequately appreciated or understood.

The preference among translators for "instant" is rather to be explained, I believe, both by the character of the argument that immediately follows the remark and by the fact that chapter 7 of the *De Sensu* contains no fewer than three references to "indivisible time".[5] The fact that Aristotle employs this expression rather than his technical term for instants (*nun*) is actually a reassurance to those who read the antecedent of "each" as "instant." For since Aristotle affirms the infinite divisibility of time in *Physics* IV.11, "indivisible time" cannot refer to a temporal atom – that is, a smallest interval of time – for he denies that there is any such thing. So he must use the expression as a stylistic variant of "now," by which he refers to an instant.[6] Further, since he denies that the now is itself any part of time, his use of "indivisible time" here in the *De Sensu* must employ a rather looser sense of "time" than is used in the formulation of his temporal theory, a looser sense that must include instants or nows. But if indivisible times – instants – are in this context counted among the class of times, then reading the antecedent of "each" as "time" will *ipso facto* include instants. Therefore, it is really quite safe to understand the remark that "it is possible to perceive each" as committing Aristotle to the perceptibility of instants.[7]

But now we face another problem, because the remark so understood is ambiguous between two rather different claims. The ambiguity is most perspicuous in a formal quantified modal language:

(T1) $\forall x(Ix \rightarrow \Diamond Pax)$
(T2) $\Diamond\forall x(Ix \rightarrow Pax)$

[4] Ross (1973, 225) draws a connection between *Sens.* 7 and *Phys.* IV.11, but he sees the justificatory relation running in the other direction – from the *Phys.* to the *Sens.*

[5] ἀτόμῳ χρόνῳ occurs at 447a14 and again at 448b20, ἄτομον χρόνον at 447b17.

[6] Cf. Aristotle's use of the phrase ἐν ἀχωρίστῳ χρόνῳ ("in an undivided time") in *An.* III.2 immediately following two sentences in which no fewer than four uses of νῦν appear. More on this passage below.

[7] In addition to the "atomic" and "instant" reading, there is a third possible reading of "indivisible time" that is consonant with my discussion of accidental compounds in Chapter 4. When used strictly, "the second-long interval" denotes a period that possesses its length essentially, and therefore cannot survive division. It is thus by convention atomic, and so an indivisible time. There is, however, no reason to suppose that this is what Aristotle has in mind in this portion of the *Sens.*

Of the two claims, (T1) is the weaker, for it maintains only that each instant is a possible object of perception for an arbitrary percipient agent *a*, whereas (T2) alleges that it is possible that *every* instant should be so perceived. It is more plausible to claim, by way of analogy, that each pie in a certain bakery is such that I might eat *it* for dessert than it would be to claim that it's possibly the case that I should eat *all* of them. The truth of the former requires only that I be not terribly fussy (which is in fact true), while the latter demands that I be capable of superhuman feats of consumption (which, alas, I am not).

Since (T1) is the more plausible claim, the principle of charity would recommend adopting it as the proper way of interpreting Aristotle's remark, unless there is independent reason to prefer the stronger alternative. As it happens, though, Aristotle is committed to a claim that is markedly stronger even than (T2):

(T3) $\Box \forall x(Ix \rightarrow Pax)$

That is, as a matter of necessity (in some suitable sense of "necessary"), any arbitrary percipient agent actually perceives every instant of time.

The purpose of this chapter is to examine Aristotle's commitment to (T3) and thereby to elucidate the meaning of the claim. In the course of doing so, we shall discover a striking incongruity between (T3) and a closely related view that Aristotle endorses even as he advocates for (T3), namely the claim that arbitrarily small sense-quality instances (e.g. patches of color) are merely potentially, not actually, perceived.

7.1 THE CONTAGION ARGUMENT

The *De Sensu* passage quoted above is situated within an inquiry concerning the possibility of simultaneous sensory discrimination of multiple, discrete sense-qualities. As he so often does, Aristotle takes up a particular view on the topic as a motivating specimen:

> Some of the writers who treat of concords assert that the sounds combined in these do not reach us simultaneously, but only appear to do so, their real successiveness being unnoticed whenever the time it involves is imperceptible. (448a21–3)[8]

Although it is unknown to whom the view in question should be attributed, the idea seems clear enough: in hearing a C major chord, for

[8] Ὃ δὲ λέγουσί τινες τῶν περὶ τὰς συμφωνίας, ὅτι οὐχ ἅμα μὲν ἀφικνοῦνται οἱ ψόφοι, φαίνονται δέ, καὶ λανθάνει, ὅταν ὁ χρόνος ᾖ ἀναίσθητος.

instance, what one hears is the C note for as brief a period as is perceptible, then E, then G, then C again, then E, and so on, for as long as the chord is sustained. By hypothesis, one doesn't hear any pair of notes simultaneously, so what's needed is an explanation for the fact that one seems to hear them simultaneously. The explanation, on the juxtapositional theory, is that the discrete episodes of perception are so small and packed together so tightly that their distinctness goes unnoticed.

Now if this is the substance of the view, it is subject to any number of objections quite unrelated to Aristotle's concern here.[9] But Aristotle rejects the juxtapositional theory of concords because he regards the very idea of imperceptible times as incoherent. His argument to this conclusion is developed in two stages. The first stage occurs at *De Sensu* 448a26–30:

> For if, [i] when someone perceives himself or another individual during a continuous period of time it cannot at that time escape his notice that he exists, but [ii] there is within the continuous time some quantum so small as to be wholly imperceptible, then it's clear that [iii] he would at that time be unaware whether he was himself and whether he saw or perceived.[10]

The larger argument of which this is a part is undoubtedly intended to be a reductio of the second premise. But what is the point of this narrower argument? Let us consider the first premise: "when someone perceives himself or another individual during a continuous period of time it cannot at that time escape his notice that he exists." It would be a mistake to think that the inclusion of the phrase "or another individual" saves the premise from triviality, because it would be far from trivial even without the phrase. The claim that perceiving oneself necessitates awareness of one's own existence is certainly not self-evidently true: simply perceiving some object *o* is quite a different matter from perceiving *that o exists*. The

[9] Here's but one such objection. Suppose that within the conditions of the scenario just described, the value of temporal duration below which perception is impossible is some definite quantity, *q*. Now assume that in hearing the C major chord, one has a series of contentful "*petites*" perceptions – C, E, G, C, E, G, etc. – in which each element of the series has a temporal duration of *q*. The experience so conceived would be phenomenologically indistinguishable from one produced by the successive *q*-long soundings of the individual notes. That is, a 3*q*-long sounding of the chord would be subjectively indistinguishable from three individual *q*-long soundings of C, E, and G. But that being so, it would be impossible to hear a C11 chord in any temporal period of less than 6*q* length, since it is composed of six notes: C, E, G, B flat, D, and F. Worse yet, on this view hearing a C11 chord would, quite implausibly, be phenomenologically indistinguishable from hearing a 3*q*-long C major chord (C, E, G) followed immediately by a B flat major chord (B flat, D, F) of the same duration.

[10] εἰ γὰρ ὅτε αὐτὸς αὑτοῦ τις αἰσθάνεται ἢ ἄλλου ἐν συνεχεῖ χρόνῳ, μὴ ἐνδέχεται τότε λανθάνειν ὅτι ἐστίν, ἔστι δέ τις ἐν τῷ συνεχεῖ καὶ τοσοῦτος ὅσος ὅλως ἀναίσθητός ἐστι, δῆλον ὅτι τότε λανθάνοι ἂν εἰ ἔστιν αὐτὸς αὑτόν, καὶ εἰ ὁρᾷ καὶ αἰσθάνεται.

former requires (it would seem) only that the object stand in the appropriate *causal* relation to the percipient agent; the latter requires an *intensional* relation between the object and percipient agent, that the agent stands in the requisite relation with a certain proposition of which the object is a constituent.[11] So even if Aristotle had omitted the phrase "or another individual," the premise would nonetheless have made a substantive claim about the nature of perception.

His inclusion of the phrase is not without significance, however, for his claim is not that some agent *A*'s perceiving object *o* entails *A*'s perceiving that *o* exists. His claim, rather, is that *A*'s perceiving *o* entails *A*'s perceiving that *A* exists.[12] Thus, if the argument employed in this chapter of the *De Sensu* is indicative of Aristotle's general theory of perception, then as he has it every ordinary act of (human) perception is accompanied by an additional reflexive, apperceptual activity, the "perception of perception" as he calls it. And, indeed, there is ample textual evidence (and sufficiently broad agreement about the evidence) to bear this interpretation.[13]

As interesting as this element of Aristotle's theory of perception might be, it contains nothing to indicate how the first premise so construed engages the second premise. What does the perceptibility of times have to do with the perception of ordinary objects, or even with the perception of perception?

Recall Aristotle's claim in *Physics* IV.11 that perception of motion is both necessary and sufficient for the perception of time:

> Now we perceive motion and time together; even if it is dark and we are not subject to bodily sensations, but some sort of motion occurs in the soul, it immediately seems that some time has also passed together with it; and not only that but also, when some time is thought to have passed, some movement also along with it seems to have taken place. (219a3–8)[14]

[11] Cf. Chapter 3, note 52.

[12] Although the present passage makes the latter claim, Aristotle is also committed to the former – see, e.g., *An.* III.3. Also cf. *NE* 1170a25 ff.

[13] See, e.g., *NE* 1170a29–b1, *Mem.* 452b26–8, *Somn.* 455a13–b12, *An.* 425b12 ff., and *Phys.* 244b10–245a2 (quoted in part below). Caston's (2002) treatment of the issue is excellent, as is Kosman's (1975). For different ways of understanding Aristotle's views on apperception but which are nonetheless in general agreement on this larger issue, see Ross (1973), 30–3; Modrak (1987), 66–7; Kahn (1992), 364–5; Everson (1997), 141–8. Annas' (1992) treatment of memory makes a similar move, according to which what she calls "personal memory" has reflexive content. More on this issue in section 8.2 below.

[14] ἅμα γὰρ κινήσεως αἰσθανόμεθα καὶ χρόνου· καὶ γὰρ ἐὰν ᾖ σκότος καὶ μηδὲν διὰ τοῦ σώματος πάσχωμεν, κίνησις δέ τις ἐν τῇ ψυχῇ ἐνῇ, εὐθὺς ἅμα δοκεῖ τις γεγονέναι καὶ χρόνος. ἀλλὰ μὴν καὶ ὅταν γε χρόνος δοκῇ γεγονέναι τις, ἅμα καὶ κίνησίς τις δοκεῖ γεγονέναι.

So on Aristotle's view, if some agent *A* perceives any movement whatsoever, *A* cannot fail to perceive some time as accompanying the movement. Now this claim is immediately relevant to the point made above about perception of perception because the latter consists, at least in part, in being aware of certain movements set up in one's body:

> [A]ctual perception is a movement through the body in the course of which the sense is affected in a certain way ... [T]he inanimate is unconscious of being affected, whereas the animate is conscious of it, though there is nothing to prevent the animate also being unconscious of it when the alteration does not concern the senses. (*Phys.* 244b11–245a2)[15]

The final qualifying clause transparently implies that whenever the alteration does concern the senses, the percipient agent will in fact be aware of the psychic movements underpinning his perceptual act.

This line of interpretation suggests that the point of this first stage of the argument against imperceptible times is to provide grounds for the following claim: if there are imperceptible times, then perception is discontinuous. The argument runs as follows:

1. If there are imperceptible times, then during any such time *t*, a percipient agent *A* will be unaware of any movement. [Concomitance of perceiving time and movement.]
2. If *A* is, at *t*, unaware of any movement, then *A* is, at *t*, unaware of any of *A*'s perceptions. [Perception as psychic *kinêseis*.]
3. If *A* is, at *t*, unaware of any of *A*'s perceptions, then *A* is not perceiving at *t*. [Ubiquity of apperception.]
4. If *A* is not perceiving at *t*, then (given that *t* could be included within a larger temporal interval in which *A* does perceive) perception is not continuous. [Analytic.]
5. Therefore, if there are imperceptible times, perception is not continuous. [From 1–4.]

Whatever other objection one might level against this argument, it is clearly insufficient to dispatch the claim that there might be imperceptible times, for there is no obvious reason to malign the idea that perception is discontinuous. (Indeed, our best current scientific evidence supports a

[15] ἡ γὰρ αἴσθησις ἡ κατ' ἐνέργειαν κίνησίς ἐστι διὰ τοῦ σώματος, πασχούσης τι τῆς αἰσθήσεως ... καὶ τὸ μὲν λανθάνει, τὸ δ' οὐ λανθάνει πάσχον. οὐδὲν δὲ κωλύει καὶ τὸ ἔμψυχον λανθάνειν, ὅταν μὴ κατὰ τὰς αἰσθήσεις γίγνηται ἡ ἀλλοίωσις. I shall have quite a bit more to say about the relation between motion and perception below.

"granular" theory of sense-perception.)[16] To provide such a reason is precisely the point of the second stage of the argument in *De Sensu* 7, which follows immediately on the heels of the first stage:

> If one perceived despite this, there would be neither time in which nor thing of which one could be conscious except thus: by being percipient during part of the time or perceiving part of the thing ... Subtract [from a line AB] a part CB in which there is no sensation; then perception in one part of this whole or of one part of it gives consciousness of the whole, which is like seeing the whole earth because one sees this particular part of it, or walking a whole year because one walks during this part of it ... The same reasoning will hold with the part AC, for perception is always in a part and of a part, and it is impossible to perceive anything in its entirety. (448b1–14)[17]

Call this the "Contagion" argument. The interval AB can be any arbitrary period of time; so if one believes that there is within AB a sub-interval in which the percipient agent has no perception, then the same reasoning may be applied to the other sub-interval. And because the length of the divided interval is perfectly arbitrary, as Aristotle clearly thinks, the procedure may be carried out indefinitely. So the implication is that, unless we deny perceptionless periods altogether, they will infect the remaining time in such a way as to leave only arbitrarily small episodes of perception punctuating longish periods of unconsciousness. Aristotle regards the absurdity of this result to be sufficiently obvious to warrant rejecting the hypothesis on which it is based and therefore takes himself to have

[16] The view current among contemporary psychologists concerning auditory perception endorses minimal temporal units of perception. The smallest operative musical subdivision has been referred to as a "temporal atom," which has subsequently been dubbed the "tatum" by Bilmes (1993) as a homage to the flurry-fingering pianist, Art Tatum. Note that endorsing this view does not commit one to the juxtapositional theory of concords.

[17] καὶ εἰ αἰσθάνεται ἔτι, οὐκ ἂν αἴη οὔτε χρόνος οὔτε πρᾶγμα οὐδὲν ὅ αἰσθάνεται ἢ ἐν ῷ, εἰ μὴ οὕτως, ὅτι ἐν τούτου τινὶ ἢ ὅτι τούτου τι ὁρᾷ ... ἀφῃρήσθω ἡ τὸ ΓΒ, ἐν ᾗ οὐκ ᾐσθάνετο. οὐκοῦν ἐν ταύτης τινὶ ἢ ταύτης τι, ὥσπερ τὴν γῆν ὁρᾷ ὅλην, ὅτι τοδὶ αὐτῆς, καὶ ἐν τῷ ἐναιυτῷ βαδίζει, ὅτι ἐν τῷδε τῷ μέρει αὐτοῦ ... ὁ δ' αὐτὸς λόγος καὶ ἐπὶ τῆς ΑΓ· ἀεὶ γὰρ ἐν τινὶ καὶ τινός, ὅλου δ' οὐκ ἔστιν αἰσθάνεσθαι. (Cf. the Eleatic argument in *GC* 1.8, 325a3.) This version of the text is endorsed by G. R. T. Ross (1973), but it is rejected by Alexander and W. D. Ross (1936). They regard the opening phrase καὶ εἰ αἰσθάνεται ἔτι as a case of dittography (the final words of the preceding sentence being καὶ εἰ ὁρᾷ καὶ αἰσθάνεται) and begin the sentence with ἔτι. In our APA conference exchange, Richard Bett endorsed the Alexander/Ross text and suggested that ἔτι introduces a second argument to the conclusion that there are no imperceptible times. I agree with his point that the construction of the opening phrase is grammatically problematic, but even if the Alexander/Ross text is correct, I deny that ἔτι need introduce a separate argument to the *same* conclusion as that of the preceding argument. Rather, we have two separate arguments whose respective conclusions are premises of a larger argument. This point is made explicit immediately below.

established the continuity of perception. By *modus tollens*, then, it follows from the first stage that there are no imperceptible times.

If we are charitable enough to grant the soundness of the "Contagion" argument, it is as it stands inadequate to establish the conclusion that instants are per se perceptible, much less that every instant is as a matter of necessity actually perceived. After all, Aristotle also regards motion as continuous, and yet he quite explicitly denies the possibility of instantaneous motion.[18] There is, moreover, a remarkable incongruity between this desired conclusion and an intimately related view that Aristotle expresses in this part of the *De Sensu*. I turn now to these problems.

7.2 AN ODD JUXTAPOSITION

The juxtapositional theory of concords, which served as the motivating specimen for the "Contagion" argument, is a close cousin of the juxtapositional theory of color, which Aristotle examines in chapter 3 of the *De Sensu*. The basis of the view is as follows:

> Firstly, white and black may be juxtaposed in such a way that by the minuteness of the division of its parts each is invisible while their product is visible, and thus color may be produced. This product can appear neither white nor black, but, since it must have some color and can have neither of the two mentioned, it must be a sort of compound and a fresh kind of hue. (439b21–7)[19]

Aristotle rejects the juxtapositional theory of color because he thinks that it entails the existence of invisible spatial magnitudes and imperceptible times,[20] and he takes the "Contagion" argument to show that neither of these consequences is possible. However, his treatment of the two consequences is importantly different.

While Aristotle rejects the juxtapositional theory of color in favor of a view according to which color is a proper mixture (*mixis*) of black and white,[21] he does not deny that the juxtapositional theory is explanatorily potent within a limited range of cases. When viewed from a distance, for example, suitably small patches of black and white properly arranged

[18] Cf. *Phys.* 234a24 ff. See also section 5.4 above.

[19] ἐνδέχεται μὲν γὰρ παρ' ἄλληλα τιθέμενα τὸ λευκὸν καὶ τὸ μέλαν, ὥσθ' ἑκάτερον μὲν εἶναι ἀόρατον διὰ σμικρότητα, τὸ δ' ἐξ ἀμφοῖν ὁρατὸν οὕτω γίγνεσθαι. τοῦτο γὰρ οὔτε λευκὸν οἷόν τε φαίνεσθαι οὔτε μέλαν· ἐπεὶ δ' ἀνάγκη μέν τι ἔχειν χρῶμα, τούτων δ' οὐδέτερον δυνατόν, ἀνάγκη μικτόν τι εἶναι καὶ εἶδός τι χρόας ἕτερον.

[20] "If we accept the hypothesis of juxtaposition, we must assume not only invisible magnitude, but also imperceptible time" (ἐπὶ μὲν οὖν τῶν παρ' ἄλληλα κειμένων ἀνάγκη ὥσπερ καὶ μέγεθος λαμβάνειν ἀόρατον, οὕτω καὶ χρόνον ἀναίσθητον) (440a21–3).

[21] *Sens.* 440b ff.; cf. *GC* 328a14.

do indeed produce the appearance of gray – but only from a suitable distance.[22] His willingness to hedge on this point is consequent upon his view that every spatial magnitude is perceptible *from some distance*.[23] That is to say, for any particular spatial magnitude *M*, there is some distance *D* beyond which *M* is not visible. The distance qualification leaves open the possibility that, situated sufficiently far from the intermingled bits of black and white, a normal human being should be incapable of discriminating the individual monochromatic bits, and thus be struck as though by a genuine compound color.[24]

Another qualification on the perceptibility of spatial magnitudes is that very small magnitudes that are constituent parts of a larger magnitude are individually perceived only potentially, and not actually so.[25] The ten-thousandth part of a grain of millet is not actually discriminated per se, even though it is part of an object that is seen in its completeness. Were the part isolated by means of separation, Aristotle fears it might "dissolve into its environment, like a drop of flavor cast into the sea."[26] Hence, there is no sense in which such spatial minutiae are per se actually perceived.

Yet another condition upon the perceptibility of spatial magnitudes is that points are invisible.[27] This follows directly from Aristotle's views that anything visible has a definite color and that color inheres exclusively in surfaces.[28] Since points neither are nor have surfaces, no point is visible.

In light of these restrictions on the perceptibility of spatial magnitudes, we should regard the twin claims in the conclusion of the "Contagion" argument as only the most tenuous of partners. For the temporal claim is that every instant is necessarily actually perceived. What could account for this profound difference?

The answer falls out of the distance qualification itself. Aristotle's view is that every magnitude is visible from some distance or other. The distance in question is clearly *spatial* distance. Spatial magnitudes are perceptible when spatially displaced from the percipient agent; indeed, that's

[22] *Sens.* 440a29–30.

[23] μὴ ἐνδέχεται μηδὲν εἶναι μέγεθος ἀόρατον, ἀλλὰ πᾶν ἔκ τινος ἀποστήματος ὁρατόν (440a27–8).

[24] Note that this particular concession to the juxtapositional theory of color does not require accepting its consequence of imperceptible times.

[25] *Sens.* 446a4–6.

[26] χωριζόμεναι δ' αἱ τηλικαῦται ὑπεροχαὶ εὐλόγως μὲν ἂν καὶ διαλύοιντο εἰς τὰ περιέχοντα, ὥσπερ καὶ ἀκαριαῖος χυμὸς εἰς τὴν θάλατταν ἐκχυθείς (446a8–10). Cf. a very similar remark at *GC* 328a27–9 regarding how a drop of wine will be "dissolved … and transformed" (λύεται … καὶ μεταβάλλει) when placed in ten thousand gallons of water.

[27] *Sens.* 448b14–17. Cf. 445b11.

[28] *Sens.* 437a6, *Sens.* 439a17–30. Cf. *Top.* 131b33–6.

the only way that they are visible, as objects held directly against the eye cannot be seen.[29]

But temporal magnitudes are not perceptible when temporally displaced from the percipient agent. Items displaced in the future are objects of expectation, whereas items displaced in the past are objects of memory – sense perception is only of objects in the immediate present.[30] One can remember the day of one's first kiss and can anticipate the moment of one's death, but neither is perceptible now.

Since temporal displacement precludes perception, if an agent ever perceives anything at all, he does so in the instantaneous present, in the now. And given the conclusion of the second stage of the "Contagion" argument – namely, that perception is continuous – it follows that whenever a human being is an actual perceiver (i.e. whenever he is awake),[31] he will perceive *in* every instant of continuous time. But by the ubiquity of apperception (the "perception of perception"), any actual perceiver will, in each instant of continuous time, perceive *that* he is perceiving. Finally, since a now just is a kinetic cut *qua* individuable by perception, an actual perceiver perceives every instant in continuous time. Since Aristotle regards the truth of the premises a matter of conceptual necessity, the conclusion is also a conceptually necessary truth: necessarily, any arbitrary percipient agent perceives every instant in time. This is the meaning of (T3) above.

Some of the more salient aspects of this view will be developed in greater detail below. But before I move on to other issues, I wish to note one interesting corollary that falls out of the account presented above and then to deflect what I take to be the most obvious objection to his position. Since individual instants are not themselves times,[32] it is not possible to perceive time, strictly speaking. Rather, one perceives the boundary of a temporal interval that includes infinitely many other instants that *were* objects of perception. Thus, in order to "perceive" time, one must make use of memory and recall having perceived those other nows. And this is indeed precisely Aristotle's stated view, as he says at *De Memoria* 449b29–30, "only those animals that perceive time remember, and the

[29] *An.* 419a12–14, 423b21–3.

[30] *Mem.* 449b10–15, *Mem.* 449b26–8. Note that in the latter passage Aristotle denies the possibility of memory τοῦ δὲ νῦν ἐν τῷ νῦν – of the now in the now. One *can* have memory of the (i.e. this) present moment at some later time, but instantaneous memory is, by his lights, conceptually impossible, for part of the very concept of memory is passage of time. More on this point in Chapter 8.

[31] *Somn.* 454a2–8. [32] Recall *Phys.* 220a18–19, 21–2.

organ of memory is that by which they perceive time."[33] The latter phrase makes it abundantly clear that Aristotle uses "perceives" only very loosely when he speaks of perceiving time.[34]

The objection is closely related to this point. In the *Physics*, Aristotle defines time in terms of number, and the relevant sense of "number" invokes perception in a nontrivial way. So it would seem that perception is logically prior to time, since the former figures in the definition of the latter. And yet, when Aristotle distinguishes perception proper from memory and anticipation, he does so in terms of time. But clearly he can't have it both ways: perception and time cannot figure in each of the other's definition, since that would make each both logically prior and posterior to the other, which is impossible.

The objection fails, because it mistakenly treats an accidental characterization of perception – one that involves reference to time – as an essential characterization.[35] But Aristotle defines the capacity to perceive in *De Anima* II.12 as "the ability to receive the sensible form without the matter,"[36] and there is no hidden reference to time here. Indeed, this objection parallels the objection that Aristotle's definition of time itself must be circular on account of the fact that "motion" (*kinêsis*) appears in the definiens. As we saw in Chapter 4, that complaint is without merit.

[33] ὅσα χρόνου αἰσθάνεται, ταῦτα μόνα τῶν ζῴων μνημονεύει, καὶ τούτῳ ᾧ αἰσθάνεται.

[34] The same must be true of motion, since, Aristotle says, no motion occurs in a now.

[35] So Aristotle's conception of perception, in contrast to that of memory, is not essentially temporal, even though it entails certain temporal consequences in concert with other doctrines Aristotle endorses.

[36] ἡ μὲν αἴσθησίς ἐστι τὸ δεκτικὸν τῶν αἰσθητῶν εἰδῶν ἄνευ τῆς ὕλης (424a18–19). More on Aristotle's theory of perception in the following chapter.

CHAPTER 8

The role of imagination

Aristotle's commitment to the instantaneous nature of perception is subject to an immediate and obvious objection. Anyone who cares to reflect on the matter will report that perception *seems* to occur not in the instantaneous now, but rather in the "specious" now, so called precisely because the present appears to us (*specio*: "to look at") to have some temporal thickness. But to the extent that perceptual content simply *consists in* being appeared to in certain ways, there is no appearance/reality distinction here, and so the way that perception seems to us really is just the way that it is – namely, temporally smeared, not instantaneous.

The objection can be expressed more pointedly. If perception were instantaneous, then (contrary to fact) certain familiar phenomena would not occur. We would not, for instance, perceive the images on a movie theater's screen as moving, but rather would see them as what they actually are: sequences of still images lasting for very brief periods of time. Because that clearly isn't the case, it must be false that perception is instantaneous, and so Aristotle's view is simply mistaken.

The objection so formulated smacks of anachronism, but there are related phenomena with which Aristotle would have been acquainted. For example, anyone who has had occasion to stroll alongside a wooden picket fence will know that looking through even quite small gaps between the boards can provide to one while walking a surprisingly comprehensive image of the space and objects lying behind the fence, and more so the faster one walks. When one stops walking, though, the image dissolves. It's still possible to see through the gaps, of course, but the visual information is radically impoverished by comparison. This is a striking phenomenon, and it seems to provide undeniable evidence against Aristotle's view. For if perception were instantaneous, the formation of the image while moving would seem absolutely inexplicable.

Aristotle directly countenances the "picket-fence phenomenon" in chapter 2 of the *De Sensu*, where he evaluates theories of vision propounded by his predecessors and begins to consider the issue concerning the perception of perception. The following passage is of particular interest:

> This theory, that sight is of the nature of fire, raises a new difficulty; for unless we suppose that it is possible for a sentient subject to see a visible object without knowing it, the eye must on this theory see itself. Why then does this not happen when the eye is at rest? The answer to our difficulty, and the cause of the idea that vision is fire, must be found in the following considerations. It is always smooth surfaces that shine in the dark, though they do not create light; and the center of the eye which men call the "black" of the eye is clearly smooth. The phenomenon occurs when the eye is moved because then the effect is as though one object became two. This is due to the rapidity of the movement, which causes the seeing subject to appear different from the object seen. (437a27–b4)[1]

This passage is dialectical, but it reveals Aristotle's sensitivity to perceptual phenomena whose existence depends on the persistence of afterimages. Moving one's eyes rapidly gives rise to a visual experience in which a single seen object appears to be more than one. This sort of visual phenomenon is relevantly like the picket-fence phenomenon insofar as the content of the visual experience seems to outstrip the content that one would expect if perception were genuinely instantaneous.[2]

So how does Aristotle explain such phenomena? By appealing to *phantasia*. "*Phantasia*" is commonly translated as "imagination," and this is not an obvious mistake given the fact that *phantasia* is "an affection which lies in our power whenever we wish (for it is possible to call up images [*ommatôn*], as those do who employ images [*eidôlopoiountes*] in arranging their ideas under a mnemonic system)" (427b18–20).[3] But even though *phantasia* is conceived by Aristotle to include the active production of

[1] ἔχει δ' ἀπορίαν τοῦτο καὶ ἑτέραν. εἰ γὰρ μὴ ἔστι λανθάνειν αἰσθανόμενον καὶ ὁρῶντα ὁρώμενόν τι, ἀνάγκη ἄρ' αὐτὸν ἑαυτὸν ὁρᾶν τὸν ὀφθαλμόν. διὰ τί οὖν ἠρεμοῦντι τοῦτ' οὐ συμβαίνει; τὰ δ' αἴτια τούτου, καὶ τῆς ἀπορίας καὶ τοῦ δοκεῖν πῦρ εἶναι τὴν ὄψιν, ἐντεῦθεν ληπτέον. τὰ γὰρ λεῖα πέφυκεν ἐν τῷ σκότει λάμπειν, οὐ μέντοι φῶς γε ποιεῖ, τοῦ δ' ὀφθαλμοῦ τὸ καλούμενον μέλαν καὶ μέσον λεῖον φαίνεται. φαίνεται δὲ τοῦτο κινουμένου τοῦ ὄμματος διὰ τὸ συμβαίνειν ὥσπερ δύο γίνεσθαι τὸ ἕν. τοῦτο δ' ἡ ταχυτὴς ποιεῖ τῆς κινήσεως, ὥστε δοκεῖν ἕτερον εἶναι τὸ ὁρῶν καὶ τὸ ὁρώμενον.

[2] Also cf. *Insomn.* 2, where Aristotle first considers the similar case of shifting one's gaze between objects of different color, and second the case of shifting gaze from the rapid movement of a river to an object at rest. In each case, the qualitative character of the subsequent visual experience is influenced by that of the former. Here he also describes in striking detail how an afterimage of the sun degrades and disappears. More on this below.

[3] τοῦτο μὲν γὰρ τὸ πάθος ἐφ' ἡμῖν ἐστίν, ὅταν βουλώμεθα (πρὸ ὀμμάτων γὰρ ἔστι ποιήσασθαι, ὥσπερ οἱ ἐν τοῖς μνημονικοῖς τιθέμενοι καὶ εἰδωλοποιοῦντες). As we shall see below, the connection between *phantasia* and memory is crucial, from the standpoint of my interpretation.

mental images, it is not limited to that activity nor to the capacity for such. As he asks shortly after the passage just quoted:

> If, then, *phantasia* is that in virtue of which an image [*phantasma*] arises for us, excluding metaphorical uses of the term, is it a single faculty or disposition relative to images, in virtue of which we discriminate and are either in error or not? (428a1–4)[4]

Although Aristotle's answer to the question is not made perfectly clear in this chapter of the *De Anima*, his considered view (as we shall see below) is that *phantasia* is indeed a unified psychic faculty. The point of immediate interest in this remark is the idea that *phantasia* involves a general ability to countenance and manipulate images, whether such images are induced by acts of perception or summoned by deliberate acts of mind. This is the dialectician's account of *phantasia*; the physicist's characterization of *phantasia* is spelled out at 428b30–429a2:[5]

> If, then, *phantasia* presents no other features than those enumerated and is what we have described, then *phantasia* must be a movement resulting from an actual exercise of perception.[6]

By specifying *phantasia*'s functional role in the *psuchê*, the dialectician invokes its form, whereas the physicist focuses on the material or physiological basis of *phantasia*. Though both characterizations are essential ingredients in saying what *phantasia* is, we shall have reason to emphasize one characterization over the other in different contexts depending on our particular interests.

For example, when our aim is to explain the picket-fence phenomenon, we will likely appeal to the physiological aspect of *phantasia*. As we have seen, Aristotle is committed to the view that visual perception is complete at every instant.[7] However, actively seeing some object sets up a sequence of movements (*kinêseis*) in one's body that themselves have perceptual content that is qualitatively determinate in precisely the same

[4] εἰ δή ἐστιν ἡ φαντασία καθ' ἣν λέγομεν φάντασμά τι ἡμῖν γίγνεσθαι καὶ μὴ εἴ τι κατὰ μεταφορὰν λέγομεν, ἆρα μία τις ἔστι τούτων δύναμις ἢ ἕξις καθ' ἃς κρίνομεν καὶ ἀληθεύομεν ἢ ψευδόμεθα;

[5] Cf. *An*. 1.1, 403a27–b2, where Aristotle says that the dialectician will define anger as the desire for retribution of wrongdoing, whereas the physicist will define it as the boiling of blood around the heart.

[6] εἰ οὖν μηθὲν μὲν ἄλλο ἔχει τὰ εἰρημένα ἢ φαντασία, τοῦτο δ' ἐστὶ τὸ λεχθέν, ἡ φαντασία ἂν εἴη κίνησις ὑπὸ τῆς αἰσθήσεως τῆς κατ' ἐνέργειαν γιγνομένη.

[7] He is committed to this view independently by virtue of counting perception as a kind of activity (*energeia*), as opposed to a kind of motion. For the κίνησις/ἐνέργεια distinction, see *Met*. 1048b, *An*. 417b2 ff.

way that the content of the perceptual act itself is determinate. Because these movements originate in the sense organ and precipitate through the appropriate channels of the body to the heart,[8] they undergo a relational change with respect to proximity to the original locus of sensation and therefore "fade" in regular and predictable ways. Movements that are very proximate to the sense organ can contribute some portion of their residual content, by means of a certain form of "back pressure," to the active content of sensation.[9] It is this fact that provides the basis for the "specious present" and therefore for the picket fence phenomenon. Visual stimuli that are particularly vivid (like the glowing heads of certain fishes and other phosphorescent objects seen in the dark) set up stronger movements that are capable of contributing greater residual content to the active content of sensation. This is why Aristotle trots out the example that he does in chapter 2 of the *De Sensu*: most people are familiar with the kind of effect that arises when glowing objects move quickly in the dark (think, for example, of the "tracer" ammunition used by the military) or, alternatively, when one moves the eyes back and forth rapidly while viewing a stationary glowing object.

It might well seem perfectly mysterious how *phantasia*, considered as a sort of causal by-product of perception, could contribute its residual content to the active content of sensation. In order to dispel the mystery, we shall have to make a foray into Aristotle's theory of perception.

8.1 *AISTHÊSIS* AS THE HEADWATERS OF *PHANTASIA*

Aristotle defines perception (*aisthêsis*) in *De Anima* II.12 as "that which is capable of taking on the sensible form without the matter."[10] The notion of taking on sensible form without matter has given rise to an exegetical cottage industry, the details of whose landscape are better left to intellectual cartographers. I can, however, describe the lay of the land in simplified terms by co-opting terminology employed by Stephen Everson.[11]

On one side of the debate are the "literalists", who regard the notion of taking on sensible form without the matter as picking out a garden-variety

[8] The textual evidence for this claim will be examined in detail below.

[9] Why not *all* of their content? Because then it should be possible to literally see objects that are no longer present. Aristotle is willing to grant that there is a sense in which we can see objects that are not present (discussed below), but this is different from the sense in which we ordinarily see objects.

[10] ἡ μὲν αἴσθησίς ἐστι τὸ δεκτικὸν τῶν αἰσθητῶν εἰδῶν ἄνευ τῆς ὕλης (424a16–18).

[11] Everson (1997), 10, *passim*.

type of change in which the sense organ literally becomes qualitatively like the perceived object. So, for example, when one looks at a ripe banana, the "eye-jelly" (*korê*) literally goes yellow; and when one smells some ripe Gorgonzola, the nasal passage itself literally becomes a bit stinky. The point of Aristotle's saying that the change occurs "without the matter," according to the literalists, is to deny that the eye takes on any of the banana, or that the nose turns into cheese. The change in question is an ordinary *material* change in the sense that the matter of the eye comes to be qualitatively like the matter of the banana in a certain respect. So the matter in question is the perceptible object's matter: to say that one receives the perceptible form of the banana without the matter is to say that none of the *banana's* matter gets into the eye when it is seen.

On the other side of the dispute are the "spiritualists," who read the notion of taking on sensible form without the matter as picking out a radically different kind of change. The principal figure in the spiritualist camp is Myles Burnyeat, who maintains that:

> [R]eceiving the warmth of a warm thing without its matter means becoming warm without really becoming warm; it means registering, noticing, or perceiving the warmth without actually becoming warm.[12]

Burnyeat appeals to Aquinas in dubbing this process of becoming aware of a perceptible quality without undergoing any physiological change a "spiritual change."[13] Thus, the spiritualist reads "without the matter" as an ellipsis for "without the *agent*'s matter undergoing any alteration."[14]

The literalist/spiritualist debate is fueled in large part by a corrupt sentence in *De Anima* II.12, the chapter in which Aristotle articulates his definition of perception. The corrupt sentence is one in which Aristotle is pondering the question why the medium of scent (air) is incapable of detecting the scent that it carries. He poses an ambiguous question: "what, then, is smelling besides [*para*] being affected in a certain way?"[15] The question's ambiguity hinges on the fact that "*para*" ("besides") can mean

[12] Burnyeat (1992), 24. [13] Burnyeat (1992), 21.

[14] There are intermediate positions between literalism and spiritualism. Victor Caston (1998, 268), for example, maintains that "to receive the form 'without the matter' is just to receive a certain *transformation* of the form, where the key aspects of that form are preserved." His "information transmission" interpretation embraces the literalist account for certain cases of perception (e.g. taste and touch), but rejects it in others. For the sake of simplicity, I shall limit my discussion to literalism and spiritualism; if an account like Caston's turns out to be correct, I believe that everything I say about *phantasia* in literalist terms will survive the modifications required to accommodate Caston's view, though demonstrating this would obviously require significant work on my part.

[15] τί οὖν ἐστι τὸ ὀσμᾶσθαι παρὰ τὸ πάσχειν τι; ἢ τὸ μὲν ὀσμᾶσθαι αι αἰσθάνεσθαι (424b16–17).

either "in addition to" or "other than." (Compare his question to this one: "What could we have for dinner besides spaghetti?" Is the speaker wondering what might *compliment* the main course, or is she wondering what might be served *instead*?)

The answer Aristotle provides does nothing to disambiguate the question because it is expressed in a sentence that is grammatically defective: "*ê to men osmasthai ai aisthanesthai*" – "probably, smelling *ai* perceiving." Two different treatments of the interposed "*ai*" have been advanced by scholars. One tack is to add a kappa, turning the meaningless "*ai*" into "*kai*," meaning "also." On this fix, Aristotle's question asks what smelling is *in addition to* being affected in the same sort of way that air is affected, and his answer is that it is *also* to perceive the scent, to be aware of it as such. This is the treatment generally preferred by literalists, for obvious reasons.

The other way of dealing with the wayward "*ai*" is to dispatch it as an instance of dittography, an artifact of weary scribesmanship in which a bit of meaningless script not present in the source text is erroneously inserted into the copy-text. On this fix, Aristotle's question asks what smelling could be *if not* being affected by scent in the way that air is, and his answer is that smelling *just is* being aware of the scent. This is the spiritualists' reading of the problematic text. Burnyeat summarizes the spiritualist view in rather vivid terms: "the physical material of animal bodies in Aristotle's world is already pregnant with consciousness, needing only to be awakened to red or warmth."[16] Insofar as perception is taken to be nothing other than a "spiritual" state of conscious awareness, there is no literal, physiological change whose occurrence renders the sense organ qualitatively similar to the object of perception. In Burnyeat's words, "there [is] no deduction from physiology to perception"[17] – or conversely, for that matter.

I have always thought that the balance of evidence favors literalism, though it appeared for some time to be an interminable debate. But Everson has recorded what seems to me an absolutely straightforward and decisive refutation of the spiritualists' position:

> Indeed, the sense organs are *essentially* such as to be affected by the proper sensibles: to be an organ of the relevant kind requires being the sort of substance which will be affected by the proper sensibles [as such] … [G]iven its material constitution the [sense] organ *must* be [qualitatively] altered. This creates two major problems for a spiritualist interpreter of Aristotle. Given that the sense

[16] Burnyeat (1992), 19. [17] Burnyeat (1992), 23.

organs are not the only bodies to be affected by the proper sensibles – so that not only eyes but other transparent things will be affected by colours and not only flesh but other bodies will be affected by temperatures, etc. – it would seem that the spiritualist will need to claim that somehow, in the case of the sense organs, the matter of which they are constituted loses its normal capacities to be affected ... Further, the spiritualist will be unable to offer any motivation for Aristotle's insistence that the sense organs should have particular types of material constitution. On the literalist's account, this can be explained because the matter of a sense organ is of a type generally to be affected in the relevant way by the proper sensibles. If one accepts that the sense organs have to take on the properties of their proper objects, then it is clear why Aristotle takes them to have the material constitutions he does.[18]

As a matter of nomological necessity, some qualitative properties are such that, when two substances standing in a determinate relation relative to the property come into contact with one another, at least one of them should undergo qualitative alteration.[19] Under standard conditions, when a warm object comes into direct contact with a cool object, it is nomologically necessary that the cool object become warmer and that the warm become cooler.[20] Such considerations are raised explicitly by Aristotle in connection with his theory of perception. As he says in *De Anima* II.11, for example:

For perception is a form of being acted upon. Hence that which an object makes actually like itself is potentially such already. This is why we have no sensation of what is just as hot, cold, hard, or soft as we are, but only of what is more so. (424a1–4)[21]

The issue is raised not only at a general level of description, but also in very specific terms in connection with the particular sense modalities. In his discussion of vision in the *De Sensu*, for instance, Aristotle says:

And it is reasonable that the interior of the eye consist of water, for water is transparent. And just as there is no vision outside [the eye] without light, the same holds within: there must be transparency. And since it is not air, it must be water. For the soul, or its perceptive faculty, does not reside on the surface of

[18] Everson (1997), 84–5.

[19] I say "at least one of" rather than "both" because Aristotle regards such reciprocal causation as being typical but not universal (see *GC* I.6 322b15). In some cases, one of the two relata is so strong that it completely overwhelms the other (cf. his remark about the drop of flavor in the sea mentioned in section 7.2 above). This fact is invoked below to explain afterimage decay.

[20] Cf. *GC* I.7, 324a2–9.

[21] τὸ γὰρ αἰσθάνεσθαι πάσχειν τι ἐστίν· ὥστε τὸ ποιοῦν οἷον αὐτὸ ἐνεργείᾳ, τοιοῦτον ἐκεῖνο ποιεῖ δυνάμει ὄν. διὸ τοῦ ὁμοίως θερμοῦ καὶ ψυχροῦ ἢ σκληροῦ καὶ μαλακοῦ οὐκ αἰσθανόμεθα, ἀλλὰ τῶν ὑπερβολῶν.

the eye, but must evidently be within; consequently, the eye must be transparent and receptive of light. (438b6–9)[22]

He addresses the relation between vision, light, and transparency in the *De Anima*, as well:

The visible, then, is color, i.e. that which overlies what is in itself visible; by "in itself" we mean not that the object is by its definition visible but that it has in itself the cause of its visibility. Every color can produce movement in that which is actually transparent, and it is its very nature to do so. (418a29–b2)[23]

To insist, as the spiritualist does, that the *korê* (or "eye jelly") does not literally become yellow when one looks at a ripe banana is to saddle Aristotle with a view that is, by his own lights, in violation of the laws of nature. Color is essentially such as to produce movement in any actually transparent stuff. Since the *korê* is transparent, it cannot fail to suffer a kind of change when exposed to color.

I have raised the issue of the spiritualist/literalist debate because its outcome affects how we understand Aristotle's conception of *phantasia*, and consequently how *phantasia* fits into his view concerning the relation between perception and time. As we saw above, Aristotle's material account of *phantasia* is as "a movement [*kinêsis*] resulting from an actual exercise of perception" (429a1–2). The spiritualist faces a daunting challenge explaining the causal sequence initiated by the perceptible object and resulting, ultimately, in a phantasm, for he insists that the intervening perceptual event involves no *kinêsis* whatever: a sense organ's taking on sensible form is nothing other than its becoming aware of a perceptible object. On the other hand, the literalist is committed to no such gap in the causal sequence, since each stage involves the literal communication of a qualitative property: the yellowness of the banana causes the air to become yellow, which causes the *korê* to become yellow, which causes something else (probably a "humor" connecting the eye to the heart)[24] to

[22] καὶ εὐλόγως τὸ ἐντός ἐστιν ὕδατος· διαφανὲς γὰρ τὸ ὕδωρ. ὁρᾶται δὲ ὥσπερ καὶ ἔξω οὐκ ἄνευ φωτός, οὕτω καὶ ἐντός· διαφανὲς ἄρα δεῖ εἶναι. καὶ ἀνάγκη ὕδωρ εἶναι, ἐπειδὴ οὐκ ἀήρ. οὐ γὰρ ἐπὶ τοῦ ἐσκχάτου ὄμματος ἡ ψυχὴ ἢ τῆς ψυχῆς τὸ αἰσθητήριόν ἐστιν, ἀλλὰ δῆλον ὅτι ἐντός· διόπερ ἀνάγκη διαφανὲς εἶναι καὶ δεκτικὸν φωτὸς τὸ ἐντὸς τοῦ ὄμματος.

[23] τὸ γὰρ ὁρατόν ἐστι χρῶμα, τοῦτο δ᾽ ἐστὶ τὸ ἐπὶ τοῦ καθ᾽ αὑτὸ ὁρατοῦ· καθ᾽ αὑτὸ δὲ οὐ τῷ λόγῳ, ἀλλ᾽ ὅτι ἐν ἑαυτῷ ἔχει τὸ αἴτιον τοῦ εἶναι ὁρατόν. πᾶν δὲ χρῶμα κινητικόν ἐστι τοῦ κατ᾽ ἐνέργειαν διαφανοῦς, καὶ τοῦτ᾽ ἔστιν αὐτοῦ ἡ φύσις.

[24] Cf. *Sens.* 438b13–17, where Aristotle cites cases of blindness induced in combatants as the result of a blow struck to the temple. Their eye remains intact, but the "passages of the eye have been severed" (ἐκτμηθῆναι τοὺς πόρους τοῦ ὄμματος). Also cf. *An.* 425b23–6. More on the heart's role in perception below.

become yellow. This last item would be the "movement resulting from an actual exercise of perception," the material basis of the phantasm.

While this account likely strikes a modern sensibility as naïve, it is clearly preferable to the spiritualist account. Furthermore, it provides the resources necessary to explain how it's possible that *phantasia* should contribute its residual content to the active content of sensation. Just as it is a matter of nomological necessity that a cool hand laid on a warm boulder results in two qualitative changes (the warming of the hand and the cooling of the boulder), so too does the contact of the *korê* with the material substrate for *phantasia* result in reciprocal effect. Situated between the external medium of color (air) and the internal substrate for *phantasia* (which I'll call "ophthalmic humor"), each of which is qualitatively similar to the object originating the causal sequence, the *korê* is bound to be affected by each. Of course, the transmission of qualitative content is far stronger in the air-to-*korê* direction than it is in the humor-to-*korê* direction, for a reason we shall examine momentarily. But that there is in fact humor-to-*korê* communication of qualitative content should be a matter of little controversy. After all, Aristotle maintains that there is *korê*-to-air communication of qualitative content, and subsequently air-to-object. In chapter 2 of the *De Insomniis*, he alleges that mirrors are sometimes permanently stained by the gaze of menstruating women.[25] Setting aside the question of what sort of evidence Aristotle might have had for this alleged phenomenon (which is one of the real howlers in the corpus), let us note his explanation:

> The cause is, as we said, that the eye is not only affected by the air but also has an effect upon it and moves it, as bright objects do (for the eye is a bright object and has color) … And the air is moved by [the bloody discharges in the eye] and has a certain effect on the air on the surface of the mirror which is continuous with it, that is it makes that air affected in the same way that it is itself; and the air on the mirror affects the surface of the mirror. (459b34–460a13)[26]

The discussion of this putative phenomenon falls immediately on the heels of an examination of two instances of the picket-fence phenomenon,[27] and more generally the persistence of visual afterimages:

[25] 459b27 ff.

[26] αἴτιον δ', ὥσπερ εἴπομεν, ὅτι οὐ μόνον πάσχει τι ἡ ὄψις ὑπὸ τοῦ ἀέρος, ἀλλὰ καὶ ποιεῖ τι καὶ κινεῖ, ὥσπερ καὶ τὰ λαμπρά· καὶ γὰρ ἡ ὄψις τῶν λαμπρῶν καὶ ἐχόντων χρῶμα … ὁ δ' ἀὴρ κινεῖται ὑπ' αὐτῶν, καὶ τὸν ἐπὶ τῶν κατόπτρων ἀέρα συνεχῆ ὄντα ποιόν τινα ποιεῖ καὶ τοιοῦτον οἷον αὐτὸς πάσχει· ὁ δὲ τοῦ κατόπτρου τὴν ἐπιφάνειαν.

[27] See note 2 above.

We must suppose that something similar takes place in the case of qualitative change; for that which is heated by something hot heats the part next to it, and this continues on in succession until the source [*archê*] is reached. So it is in the case of perception, since actual perception is a kind of qualitative change. And this is why the affection continues in the sense organs, both deep within and on the superficies, not only while they are actually engaged in perception, but also even after they have ceased to do so. (459b1–8)[28]

Given the fact that the overarching theme of the *De Insomniis* is how dreams arise from (or just are a special instance of) *phantasia*, it is very tempting to read the distinction "both deep within and on the superficies" not as distinguishing the *korê* from the surface of the eye, but rather as distinguishing something behind the eye (mentioned at *Sens.* 438b13–17) from the eye itself. The identity of "the source" (*archê*) will be discussed below.

If this is correct, then the proposed explanation of the picket-fence phenomenon looks relatively promising. Perception, properly speaking, is an activity that is complete at every instant of time, by virtue of which fact we are able to perceive instants. The fact that perception *seems* to be temporally smeared is to be explained by an appeal to the fact that the qualitative content of any instantaneous state of perception is causally determined – as a matter of nomological necessity – not only by the qualitative properties of the perceptible object "upstream" (mediated by transparency, in the case of vision), but also by the qualitative properties of the material substrate of *phantasia* "downstream." Visual afterimages are not different in kind from their active ancestors; materially, they are both qualitative states of the *korê*. Rather, afterimages differ from their active ancestors principally with respect to their etiology: active images are the product of the causal efficacy of the external transparent medium, whereas afterimages are the product of the causal efficacy of the transparent stuff within the passages connecting the eye to the heart. This is their principal difference, but they obviously also differ in their vivacity. How can this be, given the nomological principle that qualitative alteration of a patient by an agent is repaid in kind? Why do afterimages decay?

The answer, clearly, has to do with the disparate magnitudes of the causes.[29] It is because the eye is so much smaller than the typical external

[28] Ὁμοίως δ' ὑπολαβεῖν τοῦτο δεῖ καὶ ἐπ' ἀλλοιώσεως· τὸ γὰρ θερμανθὲν ὑπὸ τοῦ θερμοῦ τὸ πλησίον θερμαίνει, καὶ τοῦτο διαδίδωσιν ἕως τῆς ἀρχῆς, ὥστε καὶ ἐν ᾧ τὸ αἰσθάνεσθαι, ἐπειδή ἐστιν ἀλλοίωσίς τις ἡ κατ' ἐνέργαιαν αἴσθησις, ἀνάγκη τοῦτο συμβαίνειν. διὸ τὸ πάθος ἐστὶν οὐ μόνον ἐν αἰσθανομένοις, καὶ ἐν βάθει καὶ ἐπιπολῆς.

[29] Cf. my discussion of a particularly difficult passage in the *Mem.* in section 9.1 below. The relevant notion of size here applies to the magnitude of the material medium in which the motion is

objects of vision that its effect on their qualitative states is negligible. Remember that the causation involved is kinetic, and that smaller movements of a kind *K* (whether local movements or qualitative alterations) will be overwhelmed by larger movements of the same kind.[30] The movements within the *korê* are literally smaller than the movements of their external causes, and are therefore unlikely (though not without exception, as in the case of stained mirrors) to produce noticeable external effects. Likewise, the movements within the ophthalmic humors are literally smaller than the movements in the *korê*, and are therefore incapable of producing "upstream" effects of sufficient magnitude to sustain the afterimage indefinitely.

To illustrate the interpretation on offer, consider Aristotle's description of afterimage decay resulting from looking at the sun:

> And if, after looking at the sun or some other bright object, we shut our eyes, then, if we watch carefully, it appears in the same direct line of vision (whatever it happened to be), first of all in its own color; then it changes to red, and then to purple, until it fades to black and disappears. (459b13–18)[31]

The brilliance of the sun, understood as movements of great magnitude, excites qualitatively similar, though much smaller, movements within the eye. These, in turn, cause smaller movements, yet of the same qualitative character, in the ophthalmic humor. On closing the eyes, the movements of the ophthalmic humor are sufficiently strong to fully determine the movements within the *korê*, which is no longer subject to the causal influence of any external stimulus. The phenomenal character of the image grounded in the movement thus produced is far weaker than that of the original, because it is the product of a vastly smaller efficient cause than the sun itself. The image is thus qualitatively similar to the one experienced with eyes open, but its vivacity is greatly diminished. The movement in the *korê* returns the favor to the ophthalmic humor, which in

realized, and therefore is better expressed in terms of force rather than size. (Recall from section 6.2 that motion can be measured only along one dimension, so that the only way in which one movement can be larger than another is to be longer.) Also cf. Caston's (1998), 268, "information transmission" account: "The proportional models Aristotle speaks of, for example, do not share all of the properties represented, only a certain abstract structure – at the very least, they will differ in absolute size and presumably in the dimensions along which these proportions are exhibited as well."

30 Cf. the ship-haulers in *Phys.* VII.5: the fact that a hundred men can haul a ship one kilometer in *n*-many minutes does not entail that one man can haul the same ship one kilometer in 100*n* minutes.

31 κἂν πρὸς τὸν ἥλιον βλέψαντες ἢ ἄλλο τι λαμπρὸν μύσωμεν, παρατηρήσασι φαίνεται κατ' εὐθυωρίαν, ᾗ συμβαίνει τὴν ὄψιν ὁρᾶν, πρῶτον μὲν τοιοῦτον τὴν χρόαν, εἶτα μεταβάλλει εἰς φοινικοῦν κἄπειτα πορφυροῦν, ἕως ἂν εἰς τὴν μέλαιναν ἔλθῃ χρόαν καὶ ἀφανισθῇ.

turn re-affects the *korê*. But by now magnitude of the movement has been diminished to such an extent that the resulting image in the *korê* is not only less vivid, but also qualitatively different from the original image, and the only way in which a color can become qualitatively different is through its being supplanted by a different color.[32] Since red lies between the extremes of white and black, the afterimage will be red before it disappears. A similar process of continuous causal transaction produces an afterimage of purple (which lies between red and black), after which time the movements within the ophthalmic humor are too weak to produce any movements within the *korê*.[33] This, then, is what I meant to describe above when I spoke about the residual qualitative content of *phantasia* exercising back pressure on the active content of perception.

8.2 *PHANTASIA* AND MEMORY

Phantasia serves several important functions within Aristotle's psychological theory. We have just examined its role in active perception and how it is the basis for the "specious present" and associated perceptual phenomena. But there is another psychic function served by *phantasia* "downstream" that is directly relevant to the form of time.

In the *De Memoria*, Aristotle concludes his discussion of memory (*mnême*) with the following remarks:

> Thus we have explained what memory or remembering is: that it is a state of a mental image [*phantasmatos*], related as a likeness to that of which it is an image; and to what part of us it pertains: that it pertains to the primary sense-faculty [*tou prôtou aisthêtikou*], i.e., that with which we perceive time. (451a15–18)[34]

We shall have to address the status of the so-called primary sense-faculty in due course, but the central point of interest here is that memory consists (at least in part) in the persistence of a phantasm whose origin is an act of perception. The centrality of *phantasia* within Aristotle's account of memory is therefore quite explicit by the time he completes his presentation of it. Earlier on in the account, however, its role is less obvious.

[32] Cf. *Cat.* 4a10 ff.

[33] What is Aristotle's explanation for the fact that the decaying afterimage appears only red and purple before disappearing? The beginnings of an answer might be contained in Sorabji (1972), 301, where he explains how "red and purple … are opposed to each other in a way. For one contains much black and little white, the other the opposite." Sorabji's account is encouraging, but the details of Aristotle's answer elude my comprehension.

[34] Τί μὲν οὖν ἐστὶ μνήμη καὶ τὸ μνημονεύειν εἴρηται, ὅτι φαντάσματος, ὡς εἰκόνος οὗ φάντασμα, ἕξις, καὶ τίνος μορίου τῶν ἐν ἡμῖν, ὅτι τοῦ πρώτου αἰσθητικοῦ, καὶ ᾧ χρόνου αἰσθανόμεθα. Cf. 450a27 ff.

Consider his initial characterization of memory, which is no less revealing for its lack of emphasis on *phantasia*:

> When one has knowledge or sensation without the activity of these [thinking and perceiving], that's when one remembers. In the former case, that he learned or thought it out, and in the latter, that he heard or saw it or perceived it in some other way. For it's always the case that when one actively remembers, he says within his soul that [*hoti*] he heard, or felt, or thought this before. (449b19–24)[35]

Given the intimate connection between memory, *phantasia*, and the perception of time, I think we would do well to consider the nature of memory in some detail. The final sentence in the passage just quoted indicates that Aristotle regards the act of remembering as consisting in five elements or aspects.

First, the memory has *propositional content*, as opposed to objectual content. What one remembers, for instance, is that (*hoti*) one witnessed the falling of the Berlin Wall, rather than simply remembering the Wall itself or the event of which it was a constituent. This propositional character makes memories candidate bearers of truth-value, a fact whose significance will become clear shortly.

Second, acts of remembering are *reflexive* in that the possessor of the memory himself somehow figures in the content of the act of remembering.

Third, memory is essentially *temporal* in character, and more specifically is past-oriented. As Aristotle says at 450a19–22, "whenever one actively remembers that one has seen, heard, or learned something, one always has the additional consciousness that one did so before; but before and after are in time."[36]

Fourth, memory is *imagistic*. We saw above that Aristotle's final characterization of memory is couched principally in terms of *phantasia*, and the text that follows the passage under examination makes it clear that the "this" that one remembers is a phantasm. As he says at 450a13, "But memory, even of objects of thought, doesn't occur without a phantasm."[37]

Fifth and finally, memory's content is *intensional*. The vehicle, as we might say, of memory is a certain phantasm; but the phantasm is taken to

[35] ὅταν δ' ἄνευ τῶν ἐνεργειῶν ἔχῃ τὴν ἐπιστήμην καὶ τὴν αἴθησιν, οὕτω μέμνηται, τὸ μὲν ὅτι ἔμαθεν ἢ ἐθεώρησεν, τὸ δὲ ὅτι ἤκουσεν ἢ εἶδεν ἤ ὅ τι τοιοῦτον· ἀεὶ γὰρ ὅταν ἐνεργῇ κατὰ τὸ μνημονεύειν, οὕτως ἐν τῇ ψυχῇ λέγει ὅτι πρότερον τοῦτο ἤκουσεν ἢ ᾔσθετο ἢ ἐνόησεν.

[36] ἀεὶ γὰρ ὅταν ἐνεργῇ τῇ μνήμῃ, καθάπερ καὶ πρότερον εἴπομεν, ὅτι εἶδε τοῦτο ἢ ἤκουσεν ἢ ἔμαθε, προσαισθάνεται ὅτι πρότερον· τὸ δὲ πρότερον καὶ ὕστερον ἐν χρόνῳ ἐστίν.

[37] ἡ δὲ μνήμη καὶ ἡ τῶν νοητῶν οὐκ ἄνευ φαντάσματός ἐστιν.

be *about* something else, to stand in place of (partly by virtue of its causal history) a certain event that is represented by the phantasm. Aristotle remarks:

> Just as the picture painted on a panel is at once both a picture and a likeness, and though one and the same is both, yet they differ in being, and it is possible to think of it both as a picture and as a likeness, and we must regard the phantasm within us in us similarly both as an object of thought in itself and also in relation to something else. Considered *qua* itself, the phantasm is an object of contemplation, but considered *qua* related to something else – for instance, as that of which it is likeness [*eikôn*] – it is a reminder. (450b21–7)[38]

This notion of treating some object as a sign of something else is reminiscent of Socrates' development of the divided line in Book vi of the *Republic*: considered in themselves, concrete particulars are originals, sources of imitations (e.g. shadows, reflections, paintings, and such). Considered as images, however, they are "stepping-stones" for thought (*dianoia*) that lead the inquirer to an understanding of the eternal and supersensible forms, which are the originals of concrete particulars (509d ff.).

Aristotle's ambitions here are rather more modest than Plato's, but the use to which he puts the notion of intensionality is no less sophisticated. While the propositional character of memory makes memories candidate bearers of truth-value, the intensional nature of memory provides the correspondence conditions that form the basis for fixing truth-value.[39] The lunatic Antipheron, as Aristotle says (451a9 ff.), took certain of his phantasms to be about his past; but because the events represented as occurring by these phantasms did not in fact so occur (i.e. their causal history and their content had no essential connection to one another), his would-be memories were false.

The prospect of false memory raises an interesting question concerning the content of memory's vehicle (certain phantasms). Julia Annas advances the view that the reflexive and temporal aspects of memory are part of the content of the phantasm itself:

[38] οἷον γὰρ τὸ ἐν τῷ πίνακι γεγραμμένον καὶ ζῷόν ἐστι καὶ εἰκών, καὶ τὸ αὐτὸ καὶ ἓν τοῦτ' ἐστὶν ἄμφω, τὸ μέντοι εἶναι οὐ ταὐτὸν ἀμφοῖν, καὶ ἔστι θεωρεῖν καὶ ὡς ζῷον καὶ ὡς εἰκόνα, οὕτω καὶ τὸ ἐν ἡμῖν φάντασμα δεῖ ὑπολαβεῖν καὶ αὐτό τι καθ' αὑτὸ εἶναι θεώρημα καὶ ἄλλου φάντασμα. ᾗ μὲν οὖν καθ' αὑτό, θεώρημα ἢ φάντασμά ἐστιν, ᾗ δ' ἄλλου, οἷον εἰκὼν καὶ μνημόνευμα. As we shall see in section 9.2, anticipation also makes use of phantasms as likenesses.

[39] Note that Aristotle explicitly commits himself to the possible truth and falsity of *phantasia* in several places, including *An.* iii.3.

[A] memory of Paris, on [Aristotle's] view, is having an image which is a causal result of having seen Paris in the past, and which now represents to me *my past seeing of Paris*.[40]

Annas notes that this view, according to which the representational content of the phantasm itself is reflexive and past-oriented, is rejected by some, including Sorabji.[41] Her exegetical motivation for so enriching the representational content of the phantasm strikes me as weak.[42] But setting that issue aside, it seems to me that there is good textual reason to reject Annas' view that the representational content of a memory is intrinsically past-oriented.

At *De Memoria*. 452b23 ff. Aristotle says that an agent remembers when "the movement relative to the thing [*pragmatos*] and the movement relative to the time occur together [*hama*]."[43] Not only are these two distinct "movements" jointly sufficient for remembering, they are also individually necessary, and each can occur in absence of the other (b29–30). Now I take it that the "movement relative to the thing" is the phantasm induced by perception, whose implantation in the "indivisible and ultimate" (*atomô kai eschatô*) sense organ and persistence therein constitutes a memory after some lapse of time (451a25–30). The movement relative to time, however, is more difficult to identify. I shall endeavor to do so below, but first I note that this separate movement concerning time would be utterly redundant if the phantasm representing the event itself possessed representational content that is temporal in character, and more specifically that of a past date. (Recall Aristotle's remark at 450a19 that one always has "the additional consciousness [*prosaisthanetai*] that one did so before.") So I believe that Annas is wrong to include the temporal aspect of memory in the perceptual phantasm itself: a memory of Paris need not *intrinsically* represent a past date as such.

[40] Annas (1992), 305. Emphasis added. [41] Sorabji (1972).

[42] Briefly, Annas argues that running the view this way gives Aristotle a "unified account of memory," because remembering mathematical truths and the like doesn't seem to require any phantasm at all (1992, 305). The maneuver she makes turns the memory of such truths into a memory of having learned the truth in question. But her premise is false: Aristotle insists that "the soul never thinks without a phantasm" (διὸ οὐδέποτε νοεῖ ἄνευ φαντάσματος ἡ ψυχή) (*An.* 431a16–17), and that "even when we engage in theoretical contemplation, we must have a phantasm on which to theorize" (ὅταν τε θεωρῇ, ἀνάγκη ἅμα φάντασμά τι θεωρεῖν) (*An.* 432a8–9). Likewise, he says at *Mem.* 449b35–450a1: "Nor can we think without phantasms" (καὶ νοεῖν οὐκ ἔστιν ἄνευ φαντάσματος). Annas might believe that "there is no need to assume images of any kind" for such cognitive acts, but she would thereby be in disagreement with Aristotle.

[43] Ὅταν οὖν ἅμα ἥ τε τοῦ πράγματος γίνηται κίνησις καὶ ἡ τοῦ χρόνου, τότε τῇ μνήμῃ ἐνεργεῖ.

Immediately after telling us that the two movements can come apart from one another, Aristotle indicates that the movement relative to time is sometimes precise (*metrô*), "for example, that one did such and such the day before yesterday" (453a1).[44] But it needn't be so in order for memory to occur, for "people often say that they remember [that something occurred] but not when [it occurred], whenever they don't know the precise quantity [of time]" (453a2–4).[45] Evidently, when Aristotle claims that, "when someone actively remembers, he says within his soul that he heard, or felt, or thought this before" (449b18–24), the "before" in question needn't be any more specific than "before now."

But how does one judge that something happened before now? Aristotle provides the beginnings of an answer in the paragraph immediately before his discussion of the two movements involved in remembering:

> But the most important point is that one must know the time, whether precisely or indeterminately. There is (let it be granted) something by which one judges greater and less [time]; and it is reasonable to suppose that it's like the case of magnitudes [*megethê*]: for the mind does not think of large things at a distance by reaching out toward them, in the way that some say vision occurs (for one may think of them even if they are not there), but rather by a proportionate [*analogon*] movement. For there are in the mind similar figures and movements. (452b7–11)[46]

This passage strongly suggests that judgments of time are issued, in Aristotle's view, by the same faculty that issues judgments concerning spatial magnitudes. Magnitude (*megathos*) is listed in *De Anima* II.6 as one of the common perceptibles, along with motion, rest, number, and figure. In order to appreciate the nature of the movement in the soul that contributes the temporal component of memory, then, let us examine Aristotle's brief discussion of the common perceptibles.

44 οἷον ὅτι τρίτην ἡμέραν ὁδήποτε ἐποίησεν.

45 εἰώθασι δὲ λέγειν ὅτι μέμνηνται μέν, πότε μέντοι οὐκ ἴσασιν, ὅταν τοῦ πότε μὴ γνωρίζωσι τὸ ποσὸν μέτρῳ.

46 Τὸ δὲ μέγιστον, γνωρίζειν δεῖ τὸν χρόνον, ἢ μέτρῳ ἢ ἀορίστως. ἔστω δέ τι ᾧ κρίνει τὸν πλείω καὶ ἐλάττω· εὔλογον δ' ὥσπερ τὰ μεγέθη· νοεῖ γὰρ τὰ μεγάλα καὶ πόρρω οὐ τῷ ἀποτείνειν ἐκεῖ τὴν διάνοιαν, ὥσπερ τὴν ὄψιν φασί τινες (καὶ γὰρ μὴ ὄτων ὁμοίως νοήσεἰ), ἀλλὰ τῇ ἀνάλογον κινήσει· ἔστι γὰρ ἐν αὐτῇ τὰ ὅμοια σχήματα καὶ κινήσεις.

CHAPTER 9

Time and the common perceptibles

Aristotle's view concerning the perception of various types of magnitudes has roots in Plato's thought. In the *Protagoras*, Socrates engages the dialogue's namesake about the importance of "the art of measurement":

> I will say to them [sc. ordinary men]: "Answer me this: Do things of the same size appear to you larger when seen near at hand and smaller when seen from a distance, or not?" They would say they do. "And similarly for thickness and pluralities? And equal sounds seem louder when near at hand, softer when farther away?" They would agree. "If, then, our well-being depends upon this, doing and choosing large things, avoiding and not doing the small ones, what would seem to be our salvation in life? Would it be the art of measurement [*hê metrêtikê technê*] or the power of appearance [*hê tou phainomenou dunamis*]? While the power of appearance often makes us wander all over the place confused and regretting our actions and choices with respect to things large and small, the art of measurement, in contrast, would make the appearances lose their power by showing us the truth, would give us peace of mind firmly rooted in the truth and would save our life." (356c4–e2)[1]

This endorsement of the art of measurement appears several steps into an imagined dialectical exchange between Socrates and the common man concerning weakness of will (*akrasia*). His appeal to ordinary perceptual phenomena and associated judgments concerning relative magnitudes is intended as a kind of metaphor: the "large things" and "small things" in which he is primarily interested are of course things of greater and

[1] Ὅτε δὴ τοῦτο οὕτως ἔχει, τόδε μοι ἀποκρίνασθε, φήσω. φαίνεται ὑμῖν τῇ ὄψει τὰ αὐτὰ μεγέθη ἐγγύθεν μὲν μείζω, πόρρωθεν δὲ ἐλάττω· ἢ οὔ; – Φήσουσιν. – Καὶ τὰ παχέα καὶ τὰ πολλὰ ὡσαύτως; καὶ αἱ φωναὶ αἱ ἴσαι ἐγγύθεν μὲν μείζους, πόρρωθεν δὲ σμικρότεραι; – Φαῖεν ἄν. – Εἰ οὖν ἐν τούτῳ ἡμῖν ἦν τὸ εὖ πράττειν, ἐν τῷ τὰ μὲν μεγάλα μήκη καὶ πράττειν καὶ λαμβάνειν, τὰ δὲ σμικρὰ καὶ φεύγειν καὶ μὴ πράττειν, τίς ἂν ἡμῖν σωτηρία ἐφάνη τοῦ βίου; ἆρα ἡ μετρητικὴ τέχνη ἢ ἡ τοῦ φαινομένου δύναμις; ἢ αὕτη μὲν ἡμᾶς ἐπλάνα καὶ ἐποίει ἄνω τε καὶ κάτω πολλάκις μεταλαμβάνειν ταὐτὰ καὶ μεταμέλειν καὶ ἐν ταῖς πράξεσιν καὶ ἐν ταῖς αἱρέσεσιν τῶν μεγάλων τε καὶ σμικρῶν, ἡ δὲ μετρητικὴ ἄκυρον μὲν ἂν ἐποίησε τοῦτο τὸ φάντασμα, δηλώσασα δὲ τὸ ἀληθὲς ἡσυχίαν ἂν ἐποίησεν ἔχειν τὴν ψυχὴν μένουσαν ἐπὶ τῷ ἀληθεῖ καὶ ἔσωσεν ἂν τὸν βίον;

smaller *goodness*, not spatial magnitude. So the talk of size and thickness is a heuristic intended to get Protagoras' (and Plato's readers') intuitions oriented in the desired direction. Socrates maintains that the moral defect of *akrasia* is in fact a form of ignorance – specifically, it's a case of being ignorant of the genuinely relevant facts, the relative quantity of objective goodness possessed by the competing objects of desire.

While Plato's use of spatial perspective is metaphorical, Aristotle takes such perceptual phenomena as a topic worthy of serious philosophical investigation in themselves. His first move toward addressing the issue comes in *De Anima* II.6, where he draws two important distinctions in connection with types of perceptibles. The first distinction is between things that are perceived as such (*kath hauta*) and those that are perceived only incidentally (*kata sumbebêkos*). This distinction is grounded on facts about the features of perceptible objects that are causally efficacious relative to the function of our various perceptual faculties. So, for example, since eyes are specifically attuned to color properties (by virtue of the transparency of the *korê*), color is one of the *kath hauta* perceptibles. None of our sense organs is attuned specifically to substances, let alone to male offspring of any particular man. Thus, although Aristotle is willing to grant that one sees the son of Diares, he insists that this is a case of *kata sumbebêkos* perception: the white, moving thing that one sees just happens to be the son of Diares, and to the extent that one sees that man *as* the son of Diares, the act includes content that is contributed by something other than the activity of visual perception.[2]

The second distinction is among the *kath hauta* perceptibles themselves. Some of these are perceived by just a single sense-faculty, while others can be detected by more than one. The former Aristotle dubs "proper" (*idia*) perceptibles, the latter "common" (*koina*) perceptibles. The proper perceptibles are color, sound, scent, flavor, and, in the case of touch, "many different properties" for which there is no general term other than "tangible" (418a13–14).[3] The common perceptibles are identified in *De Anima* II.6 as motion, rest, number, shape, and magnitude (418a17–18).[4] The relevance of the common perceptibles to the hylomorphic interpretation of Aristotle's temporal theory should be plain to see. Motion is the matter of time, while perception is time's form; since Aristotle develops a view

[2] See my discussion of *de dicto* perception in Chapter 3.
[3] ὄψις χρώματος καί ἀκοὴ ψόφου καὶ γεῦσις χυμοῦ,ἡ δ' ἁφὴ πλείους μὲν ἔχει διαφοράς·
[4] κοινὰ δὲ κίνησις, ἠρεμία, ἀριθμός, σχῆμα, μέγεθος·

concerning the perception of motion quite independently of his discussion of time in the *Physics*, that view will likely illuminate the latter.

Shortly after his remarks about the movement relative to time in acts of memory (quoted in the previous chapter), Aristotle claims that "we must cognize magnitude and motion with the same faculty with which we cognize time" (450a9–10).[5] He specifies the faculty in the next sentence: "Thus it's clear that the cognition of these things belongs to the primary sense-faculty [*to proton aisthêtikon*]" (450a11–12).[6] As promised, the question concerning the identity of the primary sense-faculty will be taken up momentarily; however, it is important to note that the perception of the common perceptibles, and of time, crucially relies on phantasms. The remainder of the first sentence quoted is this: "and the phantasm is an affection of the common sense-faculty" (*tês koinês aisthêseôs*) (450a10–11).[7] Thus, the movement relative to time is also a phantasm, and we ought to get clear about its etiology and content. First, though, recall that Aristotle invokes the "indivisible and ultimate" sense-faculty in the *De Memoria* as the locus of memory-phantasms (451a25–30). Evidently, then, we have at least three faculties over and above the five ordinary sense-faculties: the primary, the common, and the ultimate (and indivisible) sense-faculties. Given their apparent centrality in the perception of motion and time, it would be good to understand the identity and function of each.

Unfortunately, the situation is complicated by several facts. First, Aristotle seemingly introduces yet another sense-faculty in the *De Somno*, which he dubs the "controlling" or "ruling" (*kurion*) sense. Second, he insists in the opening chapter of Book III of the *De Anima* that there is no sense-faculty in addition to the five ordinary senses, and yet it appears as though he is committed to an additional four. Third, he states in the same chapter that "there cannot be any special faculty for the common perceptibles" (425a14–15),[8] and yet the most obvious function of the *common* sense-faculty would be precisely the perception of the common perceptibles. Let us address these problems in reverse order.

When Aristotle says that there can be no special faculty or organ for the common perceptibles, the emphasis rightly belongs on "special" (*idion*), which I have rendered as "proper" above in connection with the proper perceptibles. What it is *to be* a common perceptible is to be perceptible by more than one of the five ordinary sense-faculties; so to suggest that there

[5] μέγεθος δ᾽ ἀναγκαῖον γνωρίζειν καὶ κίνησιν ᾧ καὶ χρόνον.
[6] ὥστε φανερὸν ὅτι τῷ πρότῳ αἰσθητικῷ τούτων ἡ γνῶσίς ἐστιν·
[7] καὶ τὸ φάντασμα τῆς κοινῆς αἰσθήσεως πάθος ἐστίν.
[8] ἀλλὰ μὴν οὐδὲ τῶν κοινῶν οἷόν τ᾽ εἶναι αἰσθητήριόν τι ἴδιον.

is a special or proper faculty for the common perceptibles is incoherent. This explains why Aristotle makes the stronger claim that there *cannot* be any such sense-faculty, rather than simply to deny that there is in fact one. Rightly understood, then, he does not with this argument preclude the existence of a sense-faculty or organ for the common perceptibles. If there is such a faculty, however, it is not one of the five ordinary sense-faculties.

But this takes us directly to the second problem, namely Aristotle's insistence that there is no sense in addition to sight, hearing, smell, taste, and touch. How can he consistently maintain this view in light of his invocation of the primary, common, ultimate, and ruling sense-faculties? Are the latter four to be variously identified with members of the other five?

No. The argument that figures at the opening of Book III of the *De Anima* is, like the sequel concerning the common perceptibles just examined, focused narrowly on faculties and organs for perceiving proper perceptibles. It is an argument to the conclusion that there are no perceptible features of reality that elude our perceptual capacities; it is not an argument to the conclusion that we possess no perceptual capacities or organs in addition to the five mentioned. He wishes to establish that there is no possible sense that we *lack*, not that we have none over and above the familiar five. So he is free to say that we have sense-faculties other than sight, hearing, smell, taste, and touch.

What, then, are we to make of the other four? One tempting interpretative strategy might involve identifying some of them with others in order to avoid what looks like a proliferation of perceptual capacities. (Why *nine* sense-faculties?) But given that "primary" and "ultimate" are plausibly antonyms ("*proton*" can be translated as "first', and "*eschaton*" could be rendered "last"), one might be more than a little pessimistic about the prospects of reducing the four to one. I suppose it is unnecessary to say that a fully respectable answer to this question would occupy the better part of another book, so I wish to provide my own solution to this puzzle by simply enumerating various claims that Aristotle makes about the faculties in question as prima facie support for my solution.

The *common* sense-faculty:

C1. Apprehends common perceptibles directly (*An.* 425a27–8).
C2. Is that by virtue of which we perceive that we see, hear, etc. (*Somn.* 455a16–17).
C3. Is common to all the organs; is one; is ruling (*kurion*); varies its modes of sensitivity with each class of perceptible object (e.g. sound and color); and is closely connected with the sense of touch (*Somn.* 455a20–3).

C4. Is affected in sleep (*Somn.* 455a26).

The *primary* sense-faculty:

P1. Perceives time (*Mem.* 450a9–10).
P2. Perceives all things (*Somn.* 455b10–11).
P3. Is paralyzed in sleep (*Somn.* 454a23, 456a20–5, 458a26–32).

The *ultimate and indivisible* sense-faculty:

U1. Is where affective states are implanted to become memories (*Mem.* 451a25–30).
U2. Is the final point of arrival for sensory modifications; is one; is a single mean; differs in modes of being (*An.* 431a17–9).

The *ruling* sense-faculty:

R1. Controls all the others (*Somn.* 455a35).
R2. Is rendered powerless in sleep, thereby rendering all other sense organs powerless (*Somn.* 455b1–3).
R3. Resides in the region about the heart in sanguineous animals, which is also the origin of bodily movement (*Somn.* 456a4–6).

As cursory as they are, I believe that these characterizations encourage the thought that the common, primary, ultimate, and ruling sense-faculty are in fact one in number. That the primary and ruling sense-faculties are one is suggested by P3 and R2. That the ruling and the common sense-faculties are one is suggested by pairs R1 and C3, and R2 and C4. That the common and ultimate sense-faculties are one is suggested by C3 and U2. By the transitivity of identity, we have before us one sense-faculty (and associated organ or physiological system) with four distinct functional aspects.

The region of the heart is the *primary* sense organ in the sense that it is the *source* of the functional capacities for the familiar five senses and therefore figures (or is implicated) in their very being.[9] A dead eye is a sightless eye, and a sightless eye is an eye in name only (*An.* 412b20). This is why without touch, whose organ is the region of the heart (*Sens.* 439a1), there can be no other sensory capacity (*An.* 435b2–3).

[9] Here we have the *archê* mentioned at *Insomn.* 459b4, quoted in section 8.1 above. The heart is also the *archê* of self-movement, in which "memories and anticipations, making use of things of this kind [thoughts and phantasms of the pleasant and painful] as likenesses, are now more and now less causes of the same changes of temperature" (μνῆμαι δὲ καὶ ἐλπίδες, οἷον εἰδώλοις χρώμεναι τοῖς τοιούτοις, ὁτὲ μὲν ἧττον ὁτὲ δὲ μᾶλλον αἰτίαι τῶν αὐτῶν εἰσιν) (*MA* 8, 702a5–7 – cf. *MA* 10).

The region of the heart is the *ultimate* sense organ in the sense that it is the final point of arrival for sensory modifications. As we saw above in our examination of *phantasia*'s role in perception, qualitative modifications of the *korê* induce further modifications deeper within the body until they finally arrive at the heart.

The region of the heart is the *ruling* sense organ in the sense that it somehow coordinates the activities of the five ordinary senses. Aristotle's explanation in the opening chapter of the *Nicomachean Ethics* of how the "most authoritative science" (*kuriôtatê epistêmê*) coordinates the goals, products, and activities of the subservient sciences is instructive here.[10] Political science employs the subservient sciences (oratory, generalship, economics, etc.), and its comprehensive end includes theirs as components. Now the function of the eyes is to see; that of the ears is to hear; and so on. Thus, the function of the ruling sense-faculty is to perceive comprehensively for the sake of its possessor's well-being.[11]

Finally, the region of the heart is the *common* sense organ in two importantly related respects.

First, it coordinates the heterogeneous products of the separate, peripheral (as we may now say) sense organs, by virtue of which process its possessor apprehends the different proper sensibles as collectively mapping onto perceptible objects in various ways. Indeed, Aristotle invokes the common sense-faculty in explaining how the physiologically disjoint perceptions of the proper perceptibles (for example, of bitterness and of yellow) are unified into a single, coherent awareness of a qualitatively complex object (specifically, of bile).[12] We should also suppose, then, that the common sense is involved in the complex perception of the white moving thing (which happens to be the son of Diares) mentioned in *De Anima* II.6. Precisely because the common sense-faculty serves these functions, its ultimate role is analogous to what Kant calls the "unity of apperception."[13] Closer to his own time, in identifying the heart as the

[10] Also cf. *Met.* Δ.1, in which Aristotle says that one sense of *archê* includes the ἀρχιτεκτονικαὶ, or "architectonic arts." The very same term appears (in a different grammatical case) in *NE* I.1 as a description of the "ruling science." So the fact that the ultimate sense organ is described at *Insomn.* 459b4 as an *archê* creates a direct link between the ultimate and ruling sense organs.

[11] Compare this with P2 under "*primary* sense-faculty" above; also see *An.* 435b20 ff.

[12] Note there's no claim being made here that perceiving bitter yellow stuff *constitutes* a perception of bile in the *de dicto* sense of perceiving employed in Chapter 3; an additional noetic contribution is required for that.

[13] Kant (1965), A105, A108.

center of sensory awareness and of thought, Aristotle preserves a tradition evident in Plato, and even in Homer.[14]

Second, by C1 above, the common sense-faculty directly apprehends the common perceptibles: motion, rest, number, figure, and magnitude. But recall that he maintains that all of the common perceptibles are perceived by motion (425a17–18). Now that we understand that the common sense-faculty trades in phantasms, which are (materially) movements set up in the body by the qualitative alteration of the peripheral sense organs, we are well situated to pursue Aristotle's justification for this remark. For when Aristotle says that "we perceive magnitude by movement," the "by" in question (carried in Greek by the dative case declension of "movement") is ambiguous: does Aristotle believe that we perceive magnitudes *by means of perceiving* movements, or that our perception of magnitudes somehow *consists in* movements? His answer, as we shall see in the following section, is "both."

Let us remind ourselves of the opening questions that spawned such speculation: what is the nature of the phantasm relative to time that figures in cases of memory? The answer to this question is to be found in determining more precisely how the primary sense-faculty cognizes magnitude, motion, and time.

9.1 SCHEMAS AND PERSPECTIVES

The *De Memoria* passage where Aristotle claims that we judge temporal intervals with the same faculty with which we judge distances is a treacherous bit of text. Although we have already seen part of the difficult paragraph (at the end of section 8.2), I wish to present the passage in full, because the interpretation I shall offer differs from the standardly received interpretation.

> But the most important point is that one must know the time, whether precisely or indeterminately. There is (let it be granted) something by which one judges greater and less [time]; and it is reasonable to suppose that it's like the case of magnitudes: for the mind does not think of large things at a distance by reaching out toward them, in the way that some say vision occurs (for one may think of them even if they are not there), but rather by a proportionate movement. For there are in the mind similar figures and movements. So when one thinks of the larger things, how will that differ from one's thinking

[14] Cf. *Tht.* 194c ff., where Plato alludes to Homer's characterization in the *Iliad* of the heart as wax, a metaphor Aristotle himself employs several times in his discussion of perception, perhaps most famously in *An.* II.12.

of the smaller things? For though they are smaller, all the internal things are, as it were, proportionate to the external. Now as we may assume within a person something proportionate to the forms of things, so we may equally well assume something else proportionate to their distances. It's as though, if one has the movement AB, BE, one makes CD; for AC and CD are proportionate. Why, then, does one make CD rather than FG? Is it not because as AC is to AB, so H is to I? These movements one therefore has together. But if one wishes to think of FG, one thinks of BE in the way one did before, but now instead of H, I, one thinks of K, L. For these are related as FA is to BA. (452b8–23)[15]

The significance of the implicit diagram depends on what we take Aristotle to be doing in this passage. G. R. T. Ross contends that:

Aristotle's sole point is to show how external [distances] and [movements] may be reproduced *in parvo* in the psychical organs. His explanation is that the internal [schemas] and [movements] are analogous to the external ones, just as the sides of a small triangle are in the same proportions as those of one any number of times larger.[16]

Ross reads "the larger things" as denoting external magnitudes and "the smaller things" as denoting the internal schemas and movements that are hypothesized to represent the external "larger things." This reading is plausible insofar as immediately after asking the question regarding the difference between a subject's thinking of larger things and his thinking of smaller things, Aristotle remarks that the internal movements are all smaller than their external causes.[17] Ross' reconstruction of the implicit diagram is shown in Figure 5.

As Ross has it, AB stands as the "psychic distance" (in a sense to be explained below) to internal object BE, whereas AC is the real distance to external object CD. Although this reconstruction differs from both Themistius' and Freudenthal's – the other influential interpretations of

[15] ἔστω δέ τι ᾧ κρίνει τὸν πλείω καὶ ἐλάττω· εὔλογον δ' ὥσπερ τὰ μεγέθη· νοεῖ γὰρ τὰ μεγάλα καὶ πόρρω οὐ τῷ ἀποτείνειν ἐκεῖ τὴν διάνοιαν, ὥσπερ τὴν ὄψιν φασί τινες (καὶ γὰρ μὴ ὄντων ὁμοίως νοήσει), ἀλλὰ τῇ ἀνάλογον κινήσει· ἔστι γὰρ ἐν αὐτῇ τὰ ὅμοια σχήματα καὶ κινήσεις. τίνι οὖν διοίσει, ὅταν τὰ μείζω νοῇ, ὅτι ἐκεῖνα νοεῖ, ἢ τὰ ἐλάττω; πάντα γὰρ τὰ ἐντὸς ἐλάττω, ὥσπερ ἀνάλογον καὶ τὰ ἐκτός. ἔστι δ' ἴσως ὥσπερ καὶ τοῖς εἴδεσιν ἀνάλογον λαβεῖν ἄλλο ἐν αὐτῷ, οὕτω καὶ τοῖς ἀποστήμασιν. ὥσπερ οὖν εἰ τὴν AB BE κινεῖται, ποιεῖ τὴν ΓΔ· ἀνάλογον γὰρ ἡ ΑΓ καὶ ἡ ΓΔ. τί οὖν μᾶλλον τὴν ΓΔ ἢ τὴν ZH ποιεῖ, ἢ ὡς ἡ ΑΓ πρὸς τὴν AB ἔχει, οὕτως ἡ τὸ Θ πρὸς τὴν I ἔχει. ταύτας οὖν ἅμα κινεῖται. ἂν δὲ τὴν ZH βούληται νοῆσαι, τὴν μὲν BE ὁμοίως μοεῖ, ἀντὶ δὲ τῶν ΘΙ τὰς ΚΛ νοεῖ. αὗται γὰρ ἔχουσιν ὡς ZA πρὸς BA.

[16] Ross (1973), 279. The bracketed words appear in Greek in Ross.

[17] Recall that this feature of Aristotle's view played an important role in my explication of how *phantasia* contributes to the "specious present" in section 8.1 above.

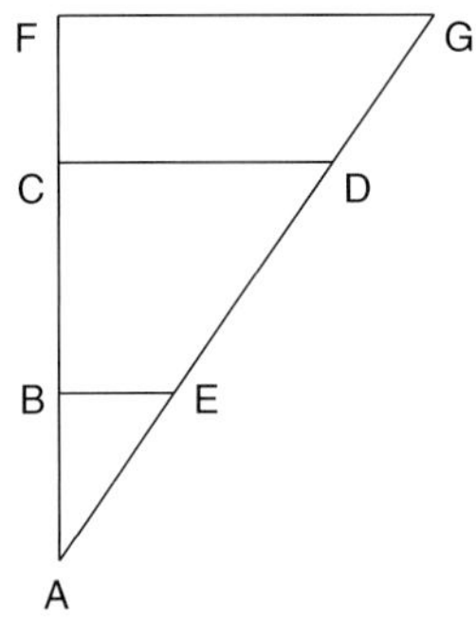

Figure 5. Ross' diagram.

this passage, both of which Ross aptly rejects[18] – the three agree that the various parts of the diagram represent respectively "outer" and "inner" movements and magnitudes, which are designated in the surrounding passage by "the larger things" and "the smaller things."

I am convinced that this reading cannot be correct. For first of all, it misplaces the level of the discussion. On Ross' view, Aristotle's question asks how one's thinking about external magnitudes differs from one's thinking about one's own psychic representations of them. This is a question far better suited for an investigation into metacognition than it is for ordinary, first-order cognition or perception. That Aristotle is concerned here with the latter is evidenced by the leading question in the paragraph: how does the mind think of large objects at a distance if not by reaching out to them (as Empedocles seems to have thought, and as Plato believes occurs in the case of vision)? Nowhere prior to the remark about all internal movements being smaller than their external causes does Aristotle suggest that these, internal and external, are the two sets of movements being distinguished from one another. So rather than treating that remark as informing the rest of the paragraph, as Ross does, I believe it is better regarded as an aside that serves to further emphasize the problem with which he is centrally concerned: *since by hypothesis* external magnitudes of different sizes are represented by internal movements that differ in size from their originals, by virtue of what fact are greater and lesser external magnitudes accurately so represented?

[18] Ross (1973), 276–83. I say "aptly" because both Themistius' and Freudenthal's diagrams require supplying alphabetic point-names not present in the original (five names in the case of Themistius' diagram, three in the case of Freudenthal's).

Thus (second), if we take the leading question within this passage to concern our ability to make discriminations among various magnitudes within the same sphere, it is entirely beside the point to read "the larger things" and "the smaller things" with Ross as designating external and internal magnitudes respectively. For on this reading the central question goes unanswered.

Ross comes closer in the following to correctly identifying the point of the passage:

> Will not an internal [schema – sc., AB] which represents a length of six feet at a certain distance [say, AC] represent one of twelve feet at double the distance [AF]? Aristotle replies that this is so, but that in the two cases we are conscious of a different proportion between the external and the internal. We have some standard by which we measure *real* size. We are conscious of the *real* distance from the eye outwards of the various objects, and hence (to state the case in modern terms) we know that an affection of the retina, which may mean a size of two inches in a near object, may mean two miles in a distant one. This is what Aristotle means when he says that [AC] is to AB in the proportion of H to I, but [AF] is to AB in the ration of K to L.[19]

Two comments are in order on Ross' answer to the question concerning the alleged indeterminacy of the intrinsic representational content of the internal schema. First, his claim that "Aristotle replies that this is so" is entirely without textual justification. Indeed, I shall argue below that Aristotle is committed to answering the question with a qualified "no." Second, far from actually answering the question Aristotle poses, Ross merely restates the implicit explanandum: we have a brute ability to be conscious of *real* size and distance. One should interpret a theory as being committed to such brute abilities only when the textual evidence is clear and univocal. But the *De Memoria* passage is neither.

The more natural way of understanding the diagram is as alternatively representing *coordinate* magnitudes within the external and internal spheres. Let us begin, then, with the external sphere. On this way of treating it, the diagram represents a viewer (at A), three perceptible objects of varying sizes (magnitudes BE, CD, and FG) displaced from the viewer at three different distances (AB, AC, and AF). Now Ross is quite correct in emphasizing Aristotle's claim that "there are in the mind similar figures and movements … proportionate to the external." So we may likewise regard the diagram as representing psychic states of our viewer, in which

[19] Ross (1973), 280. Emphasis in original.

three objects of thought are perceived or imagined as though displaced from the subject's point of view.

Perhaps the most salient feature of the diagram is its representation of the fact that the three objects will strongly coincide in terms of their apparent sizes. This fact is represented by the single angle whose vertex is A and along whose sides B, C, F, and E, D, G all lie. In contemporary terminology, the subtending angles for objects BE, CD, and FG are identical. Indeed, if the three objects were spheres, they would have identical apparent sizes and could be arranged so as to perfectly overlap from the perspective of the viewer.

So how does this begin to answer the central question of the passage – namely, how do we think of larger objects at a distance? In the imagined case of the three spheres, the only phenomenal features on which one could base judgments concerning their relative sizes and distances would be certain facts about occlusion: if sphere BE occludes sphere CD, then BE is nearer than CD and therefore the smaller of the two.[20] But what if there is only one sphere? How could an agent determine its size or distance?

The key to answering this question is Aristotle's remark that "as we may assume within a person something proportionate to the forms of things, so we may equally well assume something else proportionate to their distances." Consider the following account. I know that hippopotamuses are large creatures, because I have seen living specimens in the zoo. On Aristotle's view, I have the perceptible form of hippopotamus in my soul. Introspection reveals that my stock mental image of a hippo (one might be permitted in this context to speak of my *visual concept* of a hippo) is as of an adult specimen as though viewed from roughly ten meters. When I happen to see a mature hippo from a greater distance, it will appear proportionately smaller than my standard mental image appears to me. Or, to come at it from a different direction, if I am asked to think of a hippo at twenty meters, my mental image of the animal will be roughly half the size. Or yet again, I might be asked to think at once of a juvenile hippo at ten meters and an adult at twenty meters. In that case, the apparent size of the two imagined creatures might well be the same. What, then, would make it correct to say that I've satisfied the request, rather than to say

[20] I ignore here non-synchronous visual information (such as parallax phenomena) and proprioceptory information (such as the felt contracting and relaxing of the yoke muscles involved in binocular vision). I know of nowhere in the corpus where Aristotle acknowledges the importance – or even the existence – of such perceptual cues in depth perception.

that I've simply imagined two hippos of the same indeterminate size at an indeterminate distance?

This is precisely the question that Aristotle is pursuing in the instant passage. He begins by assuming that we have a faculty for perceiving magnitudes, and he later takes it as a premise ("*gar*" – "for," 452b13, b15) that the internal and external movements are proportionate. The question he asks is how our thinking of larger objects at a distance will differ from our thinking of smaller ones. So, again, it seems rather implausible that Aristotle would go to such lengths simply to "show how external [distances] and [movements] may be reproduced *in parvo* in the psychical organs," as Ross maintains.

Now he says, "It's as though, if one has the movement AB, BE, he makes CD; for AC and CD are proportionate." As applied to our toy example, "AB, BE" abstractly represents my standard mental image or visual concept of a mature hippo at some determinate distance, where the movements in question are phantasms that have been embedded within my primary sense organ as a memory.[21] When I am asked to think of a similar specimen at a greater distance – namely AC – I "make" CD as a superposed construction on my standard mental image. This is the point of Aristotle's remark at 452b22 that one has the movements "together" (*hama*).[22] The representational content is displayed diagrammatically in Figure 6.

The imagined beast whose distance is twice that of the nearer is imagined to be of the same size, which requires that its apparent size as imagined be half that of the nearer. This fact is represented by the relative size of the subtending angles CAD and BAE.

So in order for me to succeed in thinking of a larger object at a greater distance, something like Figure 7 represents my thought.

[21] As an aside, here is one promising way of understanding Aristotle's obscure discussion of the acquisition of *empeira* ("experience," or better: "concepts") by means of *epagôgê* ("induction") in *An. Post.* II.19. He says, "So from perception there comes memory, as we call it, and from memory (when it occurs often in connection with the same thing), a concept; for memories that are many in number form a single concept" (Ἐκ μὲν οὖν αἰσθήσεως γίγνεται μνήμη, ὥσπερ λέγομεν, ἐκ δὲ μνήμης πολλάκις τοῦ αὐτοῦ γιγνονένης ἐμπειρία· αἱ γὰρ πολλαὶ μνῆμαι τῷ ἀριθμῷ ἐμπειρία μία ἐστιν) (100a3–5 – cf. *Met.* A.1). The relevant explanatory device here is phantasms. If phantasms are movements with wave-like properties, one can see how superposition of many perceptually induced phantasms of specimens of some kind *K* can produce the concept (or more ambitiously, understanding of the universal) of *K* by means of wave interference. Constructive interference will tend to amplify the common traits among the specimens, while destructive interference will tend to diminish "noisy" features that do not properly belong to the kind in question as such.

[22] Many translators render *hama* as "simultaneously," but this is too weak. More on the importance of this term below.

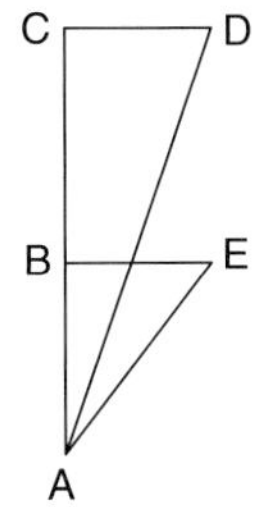

Figure 6. Two adult hippos.

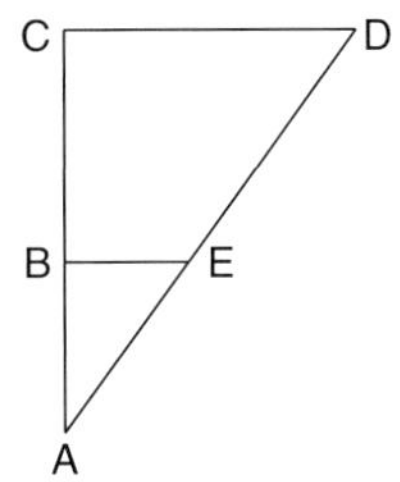

Figure 7. Juvenile and adult hippo.

Subtending angles CAD and BAE are in this case equivalent, so the apparent size of the imagined objects is the same. But, again, what sort of phenomenal content would provide the basis for thinking that this diagram accurately represents the complex thought? The answer requires that we take seriously Aristotle's remark that "as we may assume within a person something proportional to the forms of things, so we may equally well assume something else proportionate to their distances" and consider what sort of things I am supposed to be imagining. If BE is the apparent size of a juvenile hippopotamus as though viewed from ten meters, then CD might be the apparent size of an adult hippo as though viewed from twenty meters; or if BE represents an imagined golf ball as viewed at arm's length, CD might represent a tennis ball as viewed from two meters. Note that it is psychologically impossible to imagine a regulation tennis ball at arm's length together with another regulation tennis ball at five meters where both have the same apparent size. If they do have the same apparent size as imagined at different distances, then at least one of the two balls will necessarily be thought to have a size different from that of a regulation tennis ball.

So in this passage Aristotle accurately describes an interesting fact about human psychology and perception. It is a fact that has been exploited by

Hollywood under the name of "forced perspective": if a real F-16 fighter jet isn't in a film's budget, a high-quality model placed appropriately nearer the camera than the live actors can produce in the moviegoer an experience as of the real thing.

On my reconstruction, then, the movements represented by AB, BE constitute the compound phantasm serving as one's standard visual concept of a thing of kind *K* as viewed from some determinate distance *Q* and plays a role in perception analogous to the minor premise of a practical syllogism in the case of action by engaging particular episodes of perception.[23] Such a compound phantasm serves as a standard against which other phantasms (whether perceptually induced or otherwise) may be compared by means of superposition. Indeed, as we shall see below, the ability to fuse or superpose the content of distinct first-order phantasms seems to be one of the functions of *phantasia* generally. But for now, let us note that, on Aristotle's view, judgments of size and distance involve visual concepts as a matter of psychological necessity, and these are intrinsically kind-relative and have determinate content representing determinate magnitudes. But if that's right, then Ross' claim that "an internal [schema] which represents a length of six feet at a certain distance [might also] represent one of twelve feet at double the distance" is plainly false. Or, at any rate, it will be plausibly true only in cases where the visual concept is less than fully formed or is otherwise deficient, a point to be taken up momentarily. So whereas Ross reads Aristotle as being committed to a kind of brute, inexplicable ability to judge the real size and distance of external objects, I take him to have an explanation for our ability to make such judgments that is consonant with other elements of his psychological theory.

To take just one such element, this interpretation provides the theoretical resources to explain Aristotle's remark in *De Anima* III.3 that our perception of the common sensibles is liable to the greatest degree of error, "especially when the perceived object is at a distance."[24] His favorite example to illustrate error in judging size at a distance involves the sun: he says in at least three places in the corpus that the sun appears

[23] In this regard, a visual concept is closely akin to the γιγγλυμός ("ball-and-socket joint") mentioned in *An.* III.10, the same chapter in which Aristotle maintains that the sense of time – and, in particular, of the future – is a necessary psychological trait for resisting the allure of immediately present, pleasant objects. More on this below.

[24] τρίτον δὲ τῶν κοινῶν καὶ ἑπομένων τοῖς συμβεβηκόσιν οἷς ὑπάρχει τὰ ἴδια (λέγω δ' οἷον κίνησις καὶ μέγεθος) ἃ συμβέβηκε τοῖς αἰσθητοῖς· περὶ ἃ μάλιστα ἤδη ἔστιν ἀπατηθῆναι κατὰ τὴν αἴσθησιν … καὶ μάλιστα ὅταν πόρρω τὸ αἰσθητὸν ᾖ (428b22–30).

even to healthy individuals to be one foot in diameter.[25] This occurs principally on account of three facts. First, the sun is *sui generis*, so we never have occasion to experience different specimens. Second, all of our perceptual experiences of the sun are for practical purposes perfectly uniform in terms of spatial displacement: the difference between the height of the highest peak and the depth of the deepest pit is as nothing compared to the distance of the sun from any point on earth. Third, the occluding objects by which we might judge the sun's size and distance (e.g. mountains, clouds, the moon) place only a lower bound on the sun's magnitude,[26] and the celestial objects occluded by the sun (e.g. the fixed stars and the occasional planet) are of magnitude similarly difficult to determine. Thus, when it comes to the sun, there is precious little from which we might be able to construct the psychic analog of Figures 5–7.

Less egregious errors in perceiving spatial magnitudes can be explained along similar lines. For example, someone might incorrectly judge the size of a distant animal because he misjudges the animal's species; even if he correctly identifies the beast as a hippopotamus, his visual concept of a hippo (represented as "AB, BE" above) might be defective; or there might be other objects in the perceptual field generating "noisy" visual cues, as when the beast is standing next to a plant whose apparent size is for one of the foregoing reasons misjudged. On Ross' reading of the *De Memoria* passage, such errors of judgment are essentially inexplicable.

I said above that the ability to fuse or superpose the content of distinct first-order phantasms seems to be one of the functions of *phantasia* generally, and this is a point with additional consequences for my interpretation. In *De Anima* III.11, Aristotle takes up the question concerning the difference between humans and the lower animals with respect to self-movement. His answer is couched in terms of different kinds of *phantasia*:

> But how can they [imperfect (*atelês*) animals] have *phantasia*? Mustn't we say that, as their movements are indeterminate [*aoristôs*], they have *phantasia* and desire, but indeterminately? Perceptual *phantasia*, as we have said, is found in all animals, but deliberative *phantasia* only in those that are calculative [*logistikois*]; for whether one shall do this or that directly requires calculation; and

[25] *An.* 428b3–4: φαίνεται μὲν ὁ ἥλιος ποδιαῖος. Repeated with only slight variation at *Insom.* 458b28–9 and 460b17–18. Note that Aristotle is not claiming that anyone believes that the sun is only one foot in diameter, only that it appears so.

[26] Whence the so-called "moon illusion" in which the moon's apparent size is greater just above the horizon (especially a distant mountainous horizon) than it is in the open sky, even though the subtending angles are identical.

there must be a single standard to measure by, for one pursues the greater good. This requires the capacity to make a single phantasm from many. (434a4–10)[27]

That all animals have perceptual *phantasia* is a matter of nomological necessity. To be an animal is, among other things, to possess perceptive soul, and the material vehicle of *phantasia* are movements set up within the body by the qualitative alterations that occur within the peripheral sense organ during active perception. But why call this kind of *phantasia* "indeterminate"?

The answer takes its cue from the point of calling the lower animals "imperfect."[28] To say that a particular animal is imperfect in the relevant sense is to say that it is *incomplete*: it lacks something that is possessed by other animals. The missing element here is, of course, rational soul, which is possessed only by human animals. In the lower animals, the content of *phantasia* is fully determined by the content of perception and (for many species) memory, whereas humans possess an additional capacity to determine *for themselves* the content of phantasms. My view on this issue – which I acknowledge to be too controversial to be argued for convincingly here – is that such "deliberative *phantasia*" is materially identical to perceptual *phantasia* but differs with respect to its proximate cause: perceptual *phantasms* come about through acts of perception, whereas deliberative *phantasms* come to be by acts of mind (*nous*).[29] I am convinced that this is precisely the sort of (or at least one such sort of) production that Aristotle considers in *De Anima* III.5, where the infamous "maker mind" (*nous poiêtikos*) is introduced and discussed only too briefly.

At any rate, the point of calling the *phantasia* possessed by the imperfect animals "indeterminate" is just this: the content of their phantasms fails to be determined in any way *by the animal itself*, for there just isn't anything in the animal capable of doing the required work. That the content of their phantasms is qualitatively fully determinate is a matter

[27] φαντασία δὲ πῶς ἂν ἐνείη; ἢ ὥσπερ καὶ κινεῖται ἀορίστως, καὶ ταῦτ' ἔνεστι μέν, ἀορίστως δ' ἔνεστιν. ἡ μὲν οὖν αἰσθητικὴ φαντασία, ὥσπερ εἴρηται, καὶ ἐν τοῖς ἄλλοις ζῴοις ὑπάρχει, ἡ δε βουλευτικὴ [sc. φαντασία] ἐν τοῖς λογιστικοῖς (πότερον γὰρ πράξει τόδε ἢ τόδε, λογισμοῦ ἤδη ἐστὶν ἔργον· καὶ ἀνάγκη ἑνὶ μετρεῖν· τὸ μεῖζον γὰρ διώκει· ὥστε δύναται ἓν ἐκ πλειόνοων φαντασμάτων ποιεῖν). Note that the final sentence echoes the description of *epagôgê*'s role in the formation of concepts presented in *An. Post*. II.19 (cf. note 21 above).

[28] In the context of *An*. III.11, the imperfect animals under consideration are "those which have no sense but touch," namely sponges and other simple animals. The point is quite general, though.

[29] Cf. *An*. 429a5–9, where Aristotle claims that brutes act only on the basis of *phantasms* in the form of lingering perceptions "because they have no mind" (διὰ τὸ μὴ ἔχειν νοῦν).

of nomological necessity, but the source of determination is ultimately external. On the other hand, rational animals are capable of determining the content of their phantasms "from the inside" (703b18–20).[30] Recall Aristotle's characterization of *phantasia* at 427b17–20 as "an affection which lies in our power whenever we wish." Thus, the inability of lower animals to deliberate is to be explained in part by their inability to form phantasms whose content represents counterfactual situations, and (*a fortiori*) their inability to evaluate the relative goodness represented by those competing counterfactual phantasms.

As a related point, unlike other animals, human beings are capable of resisting the allure of objects of desire that are immediately present. And unlike those individuals who are weak of will, the virtuous person (more precisely, the one who possesses *sophrôsunê* – "prudence") actually *does* resist detrimental temptations, in view of their harmful consequences. The view in question consists of an appreciation of what *will* (or what would) occur *if* the power of the temptation is (or were) effective, conjoined with an evaluative judgment that such harmful consequences outweigh the benefit of yielding to the temptation, thereby delivering an all-things-considered judgment. This aspect of Aristotle's moral psychology is invoked in *De Anima* III.10, where the importance of a sense of time is made explicit:

> Appetites can conflict, and this happens when reason and desire are opposed, which occurs only in creatures with a sense of time (for while thought bids us hold back because of what is future, desire is influenced by what is immediately at hand: a pleasant object which is immediately at hand appears both unqualifiedly pleasant and unqualifiedly good, because desire cannot see the future). (433b5–10)[31]

This is Aristotle's clinical description of the art of measurement extolled by Socrates in the *Protagoras*. In my chosen terminology, it is the prescribed activity of adopting perspectives on, and thereafter evaluating, counterfactual situations. A complete understanding of Aristotle's moral theory cannot, in my view, neglect to spell out in a detailed way precisely how

[30] For an interesting (and congenial, from the standpoint of my controversial view) account of the relation between *psuchê* (and in particular, *nous*) and determination, see Kosman (1992), 357. Also cf. *MA* 11: "For thinking and imagination, as we said above, produce that which brings about the affections, since they produce the forms which bring them about" (ἡ γὰρ νόησις καὶ ἡ φαντασία, ὥσπερ εἴρηται πρότερον, τὰ ποιητικὰ τῶν παθημάτων προσφέρουσιν·).

[31] ἐπεὶ δ' ὀρέξεις γίνονται ἐναντίαι ἀλλήλαις, τοῦτο δὲ συμβαίνει ὅταν ὁ λόγος καὶ αἱ ἐπιθυμίαι ἐναντίαι ὦσι, γίνεται δ' ἐν τοῖς χρόνου αἴσθησιν ἔχουσιν (ὁ μὲν γὰρ νοῦς διὰ τὸ μέλλον ἀνθέλκειν κελεύει, ἡ δ' ἐπιθυμία διὰ τὸ ἤδη· φαίνεται γὰρ τὸ ἤδη ἡδὺ καὶ ἁπλῶς ἡδὺ καὶ ἀγαθὸν ἁπλῶς, δια τὸ μὴ ὁρᾶν τὸ μέλλον).

this activity figures into the psychological constitution and consequent life activities of the virtuous individual. Such an account must make use of a sophisticated appreciation of the function of *phantasia* within acts of moral deliberation. This project lies rather far from my present ambitions, but I need finally to show how the foregoing considerations help us to answer the question posed at the end of section 8.2: what is the nature of the phantasm relative to time that figures in cases of memory?

9.2 PHANTASMS OF PAST AND FUTURE TIMES

Active perception of oneself and one's environment induces phantasms, which propagate through various channels of the body to the central organ of perception, the heart. As we have seen, one function of the common sense-faculty is to coordinate and unify the qualitatively disparate content delivered by the peripheral sense organs and thereby to produce for its possessor a coherent and comprehensive perceptual experience of the agent himself located within an environment populated with qualitatively diverse objects. The common sense-faculty is responsible both for the "perception of perception," whereby a percipient subject is aware that she sees such-and-such, and also for fusing the disparate active content of individual senses (e.g. yellowness and bitterness) into a unified experience of a particular object or quantity of stuff (bile). As such, perceptual experience has an assertoric character: one doesn't, on Aristotle's view, simply perceive bile; rather, one perceives *that* one is perceiving something that is both yellow and bitter, which one takes to be bile. Indeed, Aristotle doesn't hesitate to use the language of propositional content in connection with the function of the common sense-faculty,[32] and we have already seen that he is committed to *phantasia*'s trading in truth-value.[33]

Under normal conditions, then, the assertoric character of perception induces in the subject complex beliefs about his environment and himself. The phantasms that lodge in the ultimate sense organ as memories retain the assertoric character of their perceptual antecedents. Thus, it often occurs that an agent, who spontaneously regards his current perceptual

[32] To cite two examples, at *An*. III.2, 425b12–13 he says that we perceive *that* (*hoti*) we see and hear (αἰσθανόμεθα ὅτι ὁρῶμεν καὶ ἀκούομεν), and just a page later at 426b20–3 claims that the common sense-faculty asserts (*legei*), thinks (*noei*), and perceives (*aisthanetai*) *that* (*hoti*) sweet differs from white (δεῖ δὲ τὸ λέγαιν ὅτι ἕτερον· ἕτερον γὰρ τὸ γλυκὺ τοῦ λευκοῦ· λέγει ἄρα τὸ αὐτό· ὥστε ὡς λέγει, οὕτω καὶ νοεῖ καὶ αἰσθάνεται).

[33] At *An*. III.3, 428a18, for example, Aristotle distinguishes *phantasia* from knowledge (*epistêmê*) and intelligence (*nous*) by pointing out the *phantasia* may be false (ἔστι γὰρ φαντασία καὶ ψευδής).

states as veridical, finds himself confronting distinct phantasmic content of diminished vivacity and distinctness, but which nonetheless possesses a similar assertoric character to that of his active perceptual states, though also less vivid. This is part of what it is to remember, on Aristotle's view.

To adapt an example from Plato's *Phaedo*: Socrates sees a lyre lying on the floor and is reminded of Simmias' musical performance the night before. Given the nature of active perception, Socrates cannot doubt (at least not without great effort) that he is looking at a lyre on the floor before him. But he is also aware of an image as of Simmias playing the lyre, an image that Socrates spontaneously regards as veridical. But Socrates can no more believe that he is currently witnessing Simmias playing the lyre than he can doubt that he is looking at an idle lyre. So Socrates spontaneously regards the event (which is itself a concrete sequence of movements and qualitative changes) represented by the memory phantasm as actual and something to which he actually stands in the perceiving relation. But not now; therefore, before.

Here, at long last, we have a purchase on the two distinct movements that figure in episodes of memory. The movement relative to the thing is simply the perceptual phantasm persisting in the ultimate sense organ. The persisting phantasm, which is itself a kind of movement, is the nomologically necessitated causal product of movements ultimately external to the subject.[34] By virtue of its causal history, it retains an assertoric character that induces assent or belief in its possessor. The movement relative to time is a separate act of the common or primary sense-faculty in which the faculty judges that the movements represented by the memory phantasm (Simmias' playing the lyre, in our example) are distinct from the movements that are the object of active perception (the lyre's lying idle). Because part of the content of the memory phantasm and the perceptual phantasm is motion, each is imbued with temporal content by virtue of figuring as objects of perception and imagination. But we know that Aristotle regards the primary sense-faculty as responsible for (i) directly perceiving the common perceptibles, including motion; (ii) superposing diverse phantasmic content; and (iii) unifying and discriminating the objects of different special sense-faculties. Therefore, in an episode of memory, the movement relative to time is materially a compound phantasm constructed by superposing the memory phantasm and the current

[34] I don't mean to suggest that the phantasm's persistence as a memory is nomologically necessary. Rather, the existence of the phantasm is, *ceteris normalis*, nomologically necessary. Whether it lodges in the ultimate sense organ and persists therein depends on contingent facts about the subject's constitution and condition. (See *Mem.* 453b1 ff.)

perceptual phantasm, and formally a judgment that the time of an event to which the agent actually stands in the perceiving relation is prior to the time of the events presently being perceived.

Several comments about this analysis are in order. First, the fundamental idea is that judgments about the past are formed by adopting perspectives on the actual. The assertoric element of a memory phantasm that places the event represented therein in one's past is precisely the residue from actively perceiving the event as it occurred. Second, the analysis is consonant with the schema appearing in *De Memoria* insofar as the judgment issued concerning displacement is founded upon the superposition of two phantasms, though the two are importantly different in at least this respect: while a visual concept might be quite stable, the standard in judgments of time is always changing. Third (however), no attempt has been made yet to say how precise (*metrô*) judgments about lapse of time are made, though I shall do so shortly. Finally, there is no suggestion that judgments regarding past events are limited to those figuring as the content of a perceptually induced phantasm. We believe all sorts of things about the remote past without having directly experienced them. So the view on offer is that our most basic sense of the past arises out of our adopting perspectives on actual perceptual phantasms. We are free to put this sense of the past in service to various constructions using deliberative *phantasia*; but the crucial point is that any phantasm as of a past event not perceived must carry with it an assertoric character comparable to that of perceptually induced memory phantasms. For if not, the event depicted will be regarded as a fiction rather than as something that actually occurred in the past.

So one actually remembers when the phantasm relative to the thing and the phantasm relative to the time are together (*hama*). How does one remember with precision (*metrô*)? Here we need to examine briefly Aristotle's account of recollection (*anamnêsis*), which provides the textual context for the very claim we wish to understand. He describes active recollection as a search (*zêtêma* – 452a8) for something preserved in memory, proceeding as a kind of inference (*sullogismos tis* – 453a10), which is why only humans are capable of recollection, though many animals have memory. The inference proceeds by the subject's inducing movements within himself (phantasms, no doubt) that bring about other movements, continuing on until the subject finds the memory he seeks (452a7–10). Many people, Aristotle remarks, successfully employ a mnemonic technique involving "places" or "loci" (*topôn* – 452a12), in which elements of the memory sequence are associated with elements of another sequence whose structure is more familiar to the subject. The discoverer of the loci method was allegedly Simonides of Ceos, who was present at a fifth-century BCE banquet in Thessaly when the roof of the hall collapsed, killing many of the diners. Simonides was enlisted to help identify the unrecognizable remains of those who had perished, and he was able to do so by imagining the seating arrangement at the tables. Proceeding from one seat to the next, Simonides could name each of the deceased diners and provide "satisfactory directions for their interment."[35] The example Aristotle provides of the loci method is perhaps less perspicuous than its inspiration:

> The reason for [the success of the loci method] is that the person passes rapidly from one thing to the next, for example from milk to white, from white to mist, and from this to moist, from which he remembers autumn, if this is the season he is trying to recall. (452a13–16)[36]

Here the target object of the search is a certain season, and the search proceeds by associations of qualitative features. Given the proximity of this passage to the one in which Aristotle claims that "the most important point is that one must know the time, whether precisely or indeterminately," it is not unreasonable to suppose that we have here an example of how one recalls the time of an event fairly precisely.

Because Aristotle offers no explicit analysis of the nature of anticipation (and little related discussion otherwise), my remarks here will be much briefer and more speculative. The point of central importance is that both memory and anticipation make use of phantasm-based perspectives. But whereas memory involves perspectives on the actual, anticipation involves perspectives on the possible, and these are induced not by external states of affairs mediated by active perception, but rather by the agent himself. Recall (from section 9.1 above) that deliberative *phantasia* is a capacity – possessed only, or perhaps only most clearly, by humans[37] – by whose activation an agent produces mental images or representations for himself. Deliberative (*bouleutikê*) *phantasia* takes its name from "*boulêsis*," or "wish," which is likewise a psychic capacity limited to fully developed

[35] Cicero (1970), 186.

[36] τὸ δ' αἴτιον ὅτι ταχὺ ἀπ' ἄλλου ἐπ' ἄλλο ἔρχονται, οἷον ἀπὸ γάλακτοςἐπὶ λευκόν, ἀπὸ λευκοῦ δ' ἐπ' ἀέρα, καὶ ἀπὸ τούτου ἐφ' ὑγρόν, ἀφ' οὗ ἐμνήσθη μετοπώρου, ταύτην ἐπιζητῶν τὴν ὥραν.

[37] Cf. *NE* 1141a27–9, where Aristotle countenances the possibility of phronetic non-human animals. More on this below.

humans.[38] Wish, Aristotle tells us, is an appetitive state whose object is an apparent good (*phainomenon agathon*).[39] Wish is very intimately related to choice (*proairesis*), which Aristotle characterizes as "deliberative desire" (*proairesis orexis bouleutikê*).[40] The distinction resides in the fact that wish takes a certain apparent good as its end, while choice defines the means to that end.[41] There is neither wish nor choice concerning things in the past, but only about what is future and contingent, things that are within our power to bring about by our own efforts.[42] Deliberate movement occurs when an agent conceives or imagines some possible good, makes it an end of her own through wish, and then searches for a sequence of intermediate causes through which she might bring about her chosen end. The last item in the sequence of causes represents the first order of business in pursuing the end, the action that is to be taken now.[43]

The role of phantasms in self-induced movement is quite evident in the *De Motu Animalium*, a treatise dedicated to that very topic:

> [T]he object we pursue or avoid in the field of action is, as has been explained, the origin of movement, and upon the thought and imagination of this there necessarily follows a heating or chilling … Blind courage and panic fears, erotic motions, and the rest of the corporeal affections, pleasant and painful, are all accompanied by heating or chilling, some in a particular member, others in the body generally. So, memories and anticipations, using things of this kind as likenesses [*eidôlois*], are now more and now less causes of the same changes of temperature. (701b33–2a7)[44]

Thus, there is a striking structural similarity between memory and recollection, on the one hand, and wish and choice, on the other. Memory consists in the preservation of a perceptually induced phantasm within the common, or ultimate, sense organ. Memories possess their assertoric character because that same aspect of perception is preserved along with the imagistic content. Recollection occurs when an agent sets out on a

[38] *NE* 1111b8. Recall that Aristotle's dialectical account of *phantasia* is "an affection which lies in our power whenever we wish" (τοῦτο μὲν γὰρ τὸ πάθος ἐφ' ἡμῖν ἐστίν, ὅταν βουλώμεθα) (427b17–18). On one natural reading of "whenever we wish," *phantasia* is a capacity that humans have the ability to activate at any time. But another reading is quite naturally suggested by the present investigation of *wish* (βουλήσις), according to which *phantasia* is a capacity that is *partly constitutive* of the act of wishing.

[39] *NE* III.4; cf. *Top.* 146b35–7a4. [40] *NE* 1111b20.

[41] *NE* 1111b26, 1113a15, 1113b3. [42] *NE* 1139b8, 1112a30. [43] *NE* 1112b19.

[44] Ἀρχὴ μὲν οὖν, ὥσπερ εἴρηται, τῆς κινήσεως τὸ ἐν τῷ πρακτῷ διωκτὸν καὶ φευκτόν· ἐξ ἀνάγκης δ' ἀκολουθεῖ τῇ νοήσει καὶ τῇ φαντασίᾳ αὐτῶν θερμότης καὶ ψύξις … θάρρη γὰρ καὶ φόβοι καὶ ἀφροδισιασμοὶ καὶ τἆλλα τὰ σωματικὰ λυπηρὰ καὶ ἡδέα τὰ μὲν κατὰ μόριον μετὰ θερμότητος ἢ ψύξεώς ἐστι, τὰ δὲ καθ' ὅλον τὸ σῶμα· μνῆμαι δὲ καὶ ἐλπίδες, οἷον εἰδώλοις χρώμεναι τοῖς τοιούτοις, ὁτὲ μὲν ἧττον ὁτὲ δὲ μᾶλλον αἰτίαι τῶν αὐτῶν εἰσίν.

search (*zêtêma* – *Mem.* 452a8) for a memory. Anticipation, on the other hand, consists in the "internal" production of a phantasm with counterfactual content. What distinguishes anticipation from a mere musing is the fact that the former, but not the latter, engages "wish" (*boulêsis*), a particular variety of appetition. Having chosen the object represented by the deliberative phantasm as an end, the agent may then search (again, *zêtêma* – *NE* 1112b20) for a sequence of causes connecting the agent's current state with the identified end.

Not every act of anticipation involves a practical action on behalf of an agent, of course. One may anticipate events that will occur as a matter of natural necessity (e.g. eclipses and solstices – cf. *GC* II.10), and also contingent events that one will play no particular role in bringing about (e.g. foreign affairs). Still, the central cases of anticipation are ones in which an agent projects his agency into the future, and these occur as has just been described. Other cases of anticipation can be constructed by imagining the event in question (an eclipse, perhaps) as being simultaneous with an action one anticipated undertaking; or one can imagine other agents anticipating undertaking future actions. Whatever the case, the point of central importance is this: Aristotle believes that our sense of the past is based upon our ability to lay down memory traces, whereas our sense of the future is based upon our ability to create ends for ourselves by means of wish. Despite the structural similarity mentioned above, these two are quite different from one another. But this is as it should be, since Aristotle regards the past and the future as also being fundamentally different from one another.

Let us now consider in the form of a fanciful illustration the view that is on the table concerning the very nature of time and judgments thereof.

CHAPTER 10

The hylomorphic interpretation illustrated

As part of a larger engineering project, Archimedes needs to determine the maximum velocity that a certain ball attains rolling down a particular incline plane. He knows that velocity is just a ratio of distance to time, *d*:*t*.[1] He also knows what distance is and how to measure it. But let us suppose that, being a bit absent-minded, Archimedes is uncertain about the nature of time and how it is measured. He remembers reading about time in the *Physics* some time ago and wonders where his copy of the text is. So he sets about trying to recall what he might have done with it.

(Commentary: Archimedes' recollection of having read the *Physics* is a psychic state concerning the past. It is founded upon a compound phantasm consisting of two perceptually induced phantasms, one of whose content is as of his reading the *Physics* (this is "the movement relative to the thing") superposed with his current perceptual phantasms. The apparent veracity of each phantasm, along with their joint inconsistency, compels Archimedes to spontaneously regard the event in question as occurring in the past.)

In attempting to determine where his copy of the text might be, Archimedes recalls that immediately after reading about time, he attended a banquet at Conon's home at which Conon mentioned Eratosthenes' interest in Aristotle's discussion of place. Visiting from Alexandria, Eratosthenes didn't have easy access to a copy of the *Physics*, so Archimedes offered to put his copy on loan. But Eratosthenes' visit was in the autumn of the year of the olive blight, and as every Syracusan knows, that was in the nineteenth year of King Hiero's rule. Archimedes remembered the time of the visit, because Eratosthenes was staying with Pausinias, whose orchard was wiped out by the blight. Archimedes now

[1] Note that, in keeping with his own view of time as a number of motion, Aristotle frequently deploys temporal values in ratios when addressing problems in natural philosophy, as he does, for example, in *Phys.* IV.8 in disproving the existence of the void (215b1 ff.).

clearly remembers delivering his copy of the *Physics* to Pausinias' house on the night of a new moon. Probably, it is there.

(Commentary: here Archimedes employs active recollection to determine both precisely when he lost possession of his book and also the conditions under which he did so. His search for the target memory phantasm proceeds by way of an initial memory phantasm's triggering other, associated phantasms, some of which are actually dated by virtue of their possessing temporal content.)

After retrieving his copy of the *Physics* from Pausinias, Archimedes reads Aristotle and is reminded that time is a certain aspect of motion, specifically its susceptibility to division into undetached parts of arbitrary length by means of perception. So he thinks to himself that in order to determine the value for *t* in his equation, he will need to define a segment of the ball's movement and compare it against some other movement that he will treat as a standard. He lays plans to conscript three assistants, to develop a chronometer of unprecedented precision, and to conduct an experiment.

(Commentary: Archimedes' psychic states about the future take the form of an adopted perspective on the possible or potential. By means of deliberative *phantasia*, Archimedes conjures phantasms as of himself conscripting assistants, and so on. Unlike memory phantasms, these have no assertoric component; however, each is informed by *boulêsis*, "wish," by virtue of which fact Archimedes attaches to the content of the phantasms a volition, "I will to (such-and-such)." Without such a component, the phantasms in question would not represent future events for Archimedes, but rather simply counterfactual situations.)

Within a month's time, Archimedes secures the human and technical resources to conduct his experiment. Implementing a series of rotating mirrors, he (quite incredibly) succeeds in developing a sundial capable of discriminating the 86,400 equal parts of the average solar day, what we today call "seconds." He decides to treat the movement of the stylus' shadow as the standard movement (which, given the causal connection between sun and shadow, is to indirectly treat the movement of the sun as standard), while the marks on the dial's face serve to define the relevant kinetic cuts against which he'll mark out the corresponding kinetic cuts in the ball's movement. He conjectures that the ball reaches maximum velocity after rolling roughly halfway down the plane. Giving two of his assistants sharp-nibbed styluses, he positions them beside the lower half of the ramp and takes his spot by the sundial. He instructs his third assistant positioned at the top of the ramp to release the ball. After the ball has passed the halfway mark, Archimedes looks to his sundial. Precisely when the shadow of the stylus arrives at one of the marks, he shouts "Now!" to instruct the first assistant to inscribe a mark on the surface of the plane at the very point where the ball makes contact with it, and precisely when the shadow arrives at the next mark on the dial's face, he shouts "Now!" to instruct the second assistant to do the same. He lays his rule on the plane and determines that ("*Eurêka!*") the maximum velocity the ball reaches on the incline plane is one and a half cubits per second.

(Commentary: the two marks that the assistants have inscribed on the incline plane – assuming for simplicity's sake that they have perfect reflexes, that their communication is instantaneous, and so on – record two kinetic cuts in the ball's movement. Perhaps the first mark on the plane records the kinetic cut specified as follows:

<!*kin*(base of the plane), exactly two cubits from the base of the plane>

whereas the second mark might record the following kinetic cut:

<!*kin*(base of the plane), exactly one half cubit from the base of the plane>.

A kinetic cut is a state of some *kinoumenon* specifiable as an ordered pair whose members are the *kinoumenon* in question and its actual location (or qualitative state, or whatever) lying on the continuum of its trajectory. (See section 5.3.) In this example, "!*kin*(base of the plane)" refers strictly to the ball insofar as it possesses the telic property *potentially being at the base of the plane*. When Archimedes says "Now!" he is himself very explicitly individuating a kinetic cut and is instructing his assistants to do so, as well.[2] Aristotle's view, recall, is that the now just is a kinetic cut's susceptibility to being individuated by an act of perception. And so when Archimedes' team records these kinetic cuts, they are quite literally marking out time, for time is "a number of motion with respect to the before and after," a measure of motion, something bounded by nows.)

10.1 SUMMARY OF PART III

The foregoing illustration brings out many important features of the hylomorphic interpretation of Aristotle's temporal theory. It does not, however, answer several questions that deserve careful treatment. Among

[2] Here I am assuming a well-defined simultaneity relation between Archimedes' shouting, etc. I shall provide analyses of various simultaneity relations in Chapter 11.

them are these three, which I present in increasing order of significance and difficulty:

1. What is the status of simultaneity on the hylomorphic interpretation?
2. What theoretical role is played by celestial motion?
3. How does the hylomorphic interpretation address the issue of temporal passage?

I take up these questions in Part IV of this book. Before turning to that task, however, it will be helpful to summarize the results of Part III.

Aristotle defines time as a number of motion. I have argued that the relevant sense of "number" involves individuating kinetic cuts and (thereby) determining units of movement. Each of these activities is effected by perception, which explains why Aristotle's discussion of time is laden with perceptual language, most particularly the text immediately surrounding his definition. The hylomorphic interpretation regards motion as the matter of time, and perception as its form. Motion is intrinsically indeterminate, but determinable, with respect to length. This determinability, or susceptibility to being divided into undetached parts of arbitrary length, is what time is. A determinate movement (or sub-movement) is *a time*, and some determinate movements are treated as tokens of a movement type which serves as a standard in timekeeping (think of the "*chous* voidings" from section 6.2).

Because *phantasia* is of central importance to his theory of perception, and because perception provides the form of time, *phantasia* also plays an essential role in Aristotle's theory of time.

First, it makes possible his claim that we are capable of perceiving individual kinetic cuts, a claim without which he would not be able to provide a hylomorphic analysis of the now as a kinetic cut *qua* perceptible. The nomologically necessary contribution of *phantasia*'s residual content to the active content of perception accounts for the specious present and explains the picket-fence phenomenon, which would otherwise count as a devastating objection to Aristotle's view that perception is instantaneous. Aristotle regards motion as one of the common perceptibles, something that is perceived immediately by the common sense-faculty. This same faculty, functioning differently in each case, is also responsible for the perception of perception, for the unity of consciousness, and for memory. If, as I argued in Chapter 3, Aristotle takes time to be an evident proper feature of motion, then the perception of time – which is also a function of the common sense-faculty – requires nothing more than the perception of motion itself and as such. This activity, I hope it is now clear, cannot fail to occur in a percipient agent whenever she is awake. For to be awake is for one's primary sense-faculty to be active, by virtue of which fact one will be aware of the goings-on in one's surroundings and in oneself.

Second, Aristotle defines memory in terms of *phantasia*, and memory has an essential temporal aspect: every memory is a memory of some event in the past. I argued that the temporal aspect of memory is not part of the proper content of the memory phantasm itself, but rather is a separate, compound phantasm generated by the primary sense-faculty's superposition of the memory phantasm's content and the subject's current perceptual phantasm. To generate such a phantasm is to adopt a perspective on the content of the target phantasm. But whereas spatial perspective employs stable visual concepts as one of the compounded phantasms, temporal perspective makes use of the ever-changing content of active perception.

Finally, I have gestured in a broad way to the idea that all varieties of perspective are founded on the activity of *phantasia*, including spatial perspectives and perspectives on future times. Indeed, Aristotle's view on the nature of time can be expressed very compactly: as distance is to space, so time is to motion. The kind of *phantasia* possessed by all animals (which Aristotle calls "perceptual *phantasia*") serves to produce perspectives on actual states of affairs countenanced directly by acts of perception. In addition to perceptual *phantasia*, human beings (and perhaps some species of nonhuman animals)[3] possess deliberative *phantasia*, a capacity to generate counterfactual phantasms and adopt perspectives on them. Perspectives on the actual provide the basis for judgments concerning the past, whereas perspectives on counterfactual situations provide the basis for judgments concerning the future. It is because lower animals lack the ability to form counterfactual phantasms (by virtue of their lacking *nous*) that they have no sense of the future. And because they have no sense of the future, they have no sense of their place in history, no motivation to undertake projects whose completion is disjoint from activities in their immediate circumstances, and therefore no culture.

[3] Cf. *NE* VI.7, 1141a27–9: "Hence practical wisdom is also attributed to some of the beasts, the ones that seem capable of forethought about their own life" (διὸ καὶ τῶν θηρίων ἔνια φρόνιμά φασιν εἶναι, ὅσα περὶ τὸν αὑτῶν βίον ἔχοντα φαίνεται δύναμιν προνοητικήν). It's unclear whether Aristotle is here simply reporting a common practice, or whether he is endorsing that practice. Aristotle does say at *PA* 648a6–7 that "bees and other similar creatures are of a more intelligent nature than many sanguineous animals" (διὸ καὶ μέλιτται καὶ ἄλλα τοιαῦτα ζῷα φρονιμώτερα τὴν φύσιν ἐστὶν ἐναίμων πολλῶν) (cf. *PA* 650b25).

PART IV

Simultaneity and temporal passage

CHAPTER 11

Simultaneity and other temporal relations

Now that we have examined both the matter and the form of time, the hylomorphic analysis is technically complete. However, there are residual issues of importance that need to be addressed. To my knowledge, no commentator of Aristotle has attempted to analyze his notion of simultaneity,[1] but the hylomorphic interpretation has ready to hand an obvious and effective analysis of this temporal relation. The analysis points up the importance of celestial motion within the theory, and we shall have to examine some of Aristotle's remarks in the *Physics* and other treatises on natural philosophy to fully appreciate the status and role of celestial motion in his account.

The most daunting residual issue concerns Aristotle's treatment of temporal passage and related matters about the present or "now." There is substantive disagreement over the question of whether Aristotle treats time as being essentially static (as it is in McTaggart's *B* series) or dynamic (as in his *A* series).[2] I shall prepare the ground for the hylomorphic view on this issue by examining the interpretations of Richard Sorabji and Fred Miller, both of whom argue that Aristotle fails to integrate temporal passage into his theory in a satisfactory way. Taking this line of argument considerably further, Norman Kretzmann claims that Aristotle, like McTaggart, believes that there is a contradiction inherent in the notion of temporal passage. The hylomorphic interpretation is inconsistent with all of these views. Specifically, I shall argue in Chapter 12 that not only is Aristotle's conception of time clearly a version of the *A* series, but it also integrates temporal passage in such a way as to be immune from one of the most devastating lines of argument against its very possibility.

[1] As we shall see below, Coope (2005) considers the possibility of such an analysis, but denies that Aristotle has the requisite conceptual resources to offer one.

[2] Noteworthies on the side of a dynamic interpretation are Ross (1936), Owen (1976), Inwood (1991), and Coope (2005). The only static interpretation other than Kretzmann (1976) that has received much attention in the literature is Wieland (1962).

In Chapter 13, we shall see how the hylomorphic interpretation dissolves the puzzles Aristotle raises in *Physics* IV.10 but does not subsequently address. Chapter 14 summarizes the interpretation developed in this book and offers a final comparison of Aristotle's and Plato's views on time. In addition, I reflect very briefly on whether Aristotle's account might inform work being done in contemporary metaphysics of time.

On Aristotle's view, a particular movement of any duration whatsoever contains infinitely many kinetic cuts. This commitment is essential for the success of the hylomorphic interpretation. Since kinetic cuts are the matter of nows, the continuity of kinetic cuts provides the material ground for the continuity of temporal intervals. However, if nows are strictly bound to individual kinetic cuts, Aristotle faces an obvious and seemingly devastating problem: every moving object would have its own now and, consequently, its own time. Recall, however, that Aristotle explicitly rejects this in one of his arguments in *Physics* IV.10 against identifying time with motion. He says that whereas a movement is only with the thing that is moving, time is everywhere equally.[3] On what grounds can he make this claim?

I note here a fundamental difference between the hylomorphic interpretation and the interpretation offered by Coope, who regards any appearance that Aristotle seeks "to explain temporal asymmetry in terms of some other, more basic kind of asymmetry" as "largely deceptive."[4] Because she rejects all so-called "reductive" interpretations of Aristotle's temporal theory, Coope is bound to attribute to Aristotle a view according to which certain temporal properties and relations are metaphysically primitive, including the relation of simultaneity:

> Aristotle's account turns out to have little in common with modern reductive explanations of temporal order. That this is so is brought out, among other things, by his readiness to take for granted the notion of simultaneity. He thinks it important to argue that simultaneous changes are all at one and the same time, but he sees no need to explain in virtue of what certain changes are simultaneous. He simply assumes the existence of this relation between changes, as if it is something basic and inexplicable.[5]

[3] *Phys.* 218b10–13. Aristotle's considered view is that time exists everywhere that motion and change (and by extension, rest) are possible. This, he notes in *Cael.* I.9, includes all of the heaven (in the third sense he distinguishes), outside of which there is neither place, void, movement, nor time (279a12–18).

[4] Coope (2005), 4.

[5] Coope (2005), 4. Cf. p. 116: "He never answers, or even raises, the question that would be central if he were giving a reductive account of temporal order. He never, that is, explains in virtue of what certain change-parts are all bounded by one and the same now."

Coope is correct in saying that Aristotle doesn't anywhere provide an explicit analysis of simultaneity, and so this presents a pointed challenge for the hylomorphic interpretation: to the extent that it is a variety of "reductive" account, it needs to provide a straightforward analysis of simultaneity, even though Aristotle himself did not articulate that analysis.

As a first step toward meeting the challenge, let us examine another passage in which Aristotle raises issues surrounding simultaneity:

> But it's possible that some other thing has been moved now, as well; so time would be a number of either of the movements. Does it follow that there's a different time, and will there be two equivalent times simultaneously, or not? [It does not follow.] For a time that is equivalent [*isos*] and simultaneous [*hama*] is one and the same; but even those that aren't simultaneous are one in form. For if there were dogs and horses, seven of each, the number would be the same. In the same way, too, time is the same number of movements that are bounded simultaneously, though one movement may be quick while the other isn't, and one may be locomotion while the other is qualitative change. (223b1–8)[6]

Aristotle draws an interesting distinction in this passage between times that are "one and the same" and times that are "one in form." I shall address this distinction soon enough, but I wish first to point out that both this translation and Coope's challenge take advantage of contextual cues in rendering "*hama*" as "simultaneous." There is of course nothing objectionable in this, but we ought not neglect the fact that the core meaning of "*hama*" is simply "together."[7] Movements that are together in time are undoubtedly simultaneous. But to address Coope's challenge squarely, we must answer this question: is there some non-temporal sense of "*hama*" that could explain the temporal relation of simultaneity?

Yes. As we saw in Chapter 8, Aristotle maintains in the *De Memoria* that an agent remembers when "the movement relative to the thing and the movement relative to the time occur together [*hama*]." I went on to argue that "*hama*" cannot in this context simply mean "simultaneously," but rather requires that the phantasms in question be fused or superposed with one another. The reason for this was that his account portrays recollection as a kind of perspective. The most familiar type of perspective is

[6] ἀλλ' ἔστι νῦν κεκινῆσθαι καὶ ἄλλο· ὧν ἑκατέρας τῆς κινήσεως εἴη ἂν ἀριθμός. ἕτερος οὖν χρόνος ἔστιν, καὶ ἅμα δύο ἴσοι χρόνοι ἂν εἶεν· ἢ οὔ; ὁ αὐτὸς γὰρ χρόνος καὶ εἷς ὁ ἴσος καὶ ἅμα· εἴδει δὲ καὶ οἱ μὴ ἅμα· εἰ γὰρ εἶεν κύνες, οἱ δ' ἵπποι, ἑκάτεροι δ' ἑπτά, ὁ αὐτὸς ἀριθμός. οὕτω δὲ καὶ τῶν κινήσεων τῶν ἅμα περαινομένων ὁ αὐτὸς χρόνος, ἀλλ' ἡ μὲν ταχεῖα ἴσως ἡ δ' οὔ, καὶ ἡ μὲν φορὰ ἡ δ' ἀλλοίωσις·

[7] The temptation to translate ἅμα as "simultaneous" can be overwhelming. For example, Ackrill's translation of the *Cat.* renders ἅμα as "simultaneous" in chapter 13, with the incredible result that the double and the half are simultaneous.

spatial perspective, and it clearly *must* involve the superposition or fusion of two phantasms. If an agent wishes to judge the distance of an object of known magnitude, it will be of no use to have a memory phantasm (what I called a "visual concept") and perceptual phantasm at the same time unless they are somehow unified in the agent's consciousness. Recall that one function of the primary sense-faculty is the unification of heterogeneous perceptual content, as well as the unification of perceptual content with other, non-perceptual content. Judgments about the past are grounded in perspectives on memory phantasms, whereas judgments about the future are grounded in perspectives on deliberative phantasms.[8] In each instance, an active perceptual phantasm constitutes part of the perspective, and it is precisely the possibility of such phantasms that provides the form for the now. Of relevance here is a lengthy passage from the *De Anima*, in which Aristotle discusses the power of the common sense-faculty to make discriminations between heterogeneous perceptible qualities:

> That which asserts the difference must be one, for sweet differs from white. It is the same thing, then, that asserts this; and as it asserts, so it thinks and perceives. It is therefore clear that it is not possible to discriminate two separate objects by means of two separate faculties. And that it is also not possible to do this in separated time is established by the following. For as what asserts the difference between the good thing and the bad thing is one and the same, so also *when* it asserts the one to be different from the other is not merely accidental to the assertion (as it is, for instance, when I now assert a difference but do not assert that there is now a difference). It asserts thus: both now and that the difference is now, and so [the judgments concerning the good thing and the bad thing must be made] together [*hama ara*]. So the discriminating sense must be undivided and must judge in an undivided time. (426b20–30)[9]

My translation "and so [the judgments concerning the good thing and the bad thing must be made] together" is an interpolation of "and so together" (*hama ara*). Most translators render this "*hama*" as

[8] Which, again, is not to say that only those states about which one might deliberate can figure in judgments concerning the future. "Deliberative *phantasia*" names our capacity to induce counterfactual phantasms which can serve as elements of a future-oriented compound phantasm when fused with the appropriate doxastic or appetitive attitude, and an active perceptual phantasm. See section 9.2.

[9] δεῖ δὲ τὸ ἓν λέγειν ὅτι ἕτερον· ἕτερον γὰρ τὸ γλυκὺ τοῦ λευκοῦ· λέγει ἄρα τὸ αὐτό· ὥστε ὡς λέγει, οὕτω καὶ νοεῖ καὶ αἰσθάνεται ὅτι μὲν οὖν οὐχ οἷόν τε κεχωρισμένοις κρινειν τὰ κεχωρισμένα, δῆλον· ὅτι δ᾽ οὐδ᾽ ἐν κεχωρισμένῳ χρόνῳ, ἐντεῦθεν· ὥσπερ γὰρ τὸ αὐτὸ λέγει ὅτι ἕτερον τὸ ἀγαθὸν καὶ τὸ κακόν, οὕτω καὶ ὅτε θάτερον λέγει ὅτι ἕτερον καὶ θάτερον, οὐ κατὰ συμβεβηκὸς τὸ ὅτε (λέγω δ᾽, οἷον νῦν λέγω ὅτι ἕτερον, οὐ μέντοι ὅτι νῦν ἕτερον), ἀλλ᾽ οὕτω λέγει, καὶ νῦν καὶ ὅτι νῦν· ἅμα ἄρα. ὥστε ἀχώριστον καὶ ἐν ἀχωρίστῳ χρόνῳ.

"simultaneously," which is certainly part of the conclusion that Aristotle is laboring to establish. But the central point of this passage is precisely that judgments about difference must be unitary – issued by a single faculty in a single act.

It is within the power of the common sense-faculty to unify and discriminate various perceptual phantasms, to make its possessor aware of the fact that she is perceiving, and to directly perceive motion. A logically consequent power, then, discriminates various perceived movements, including the agent's own psychic movements. When such perceptions are appropriately together (*hama*), the agent has an awareness of a unitary present that includes diverse movements – she perceives herself as now being immersed in an environment populated with moving and mobile substances, of which she herself is one.[10]

11.1 TWO FUNDAMENTAL TEMPORAL NOTIONS

These considerations about the unifying power of the common sense-faculty form the basis for a fundamental notion of the simultaneity relation, one that obtains between kinetic cuts:

Strict simultaneity: kinetic cut K_1 is strictly simultaneous with kinetic cut K_2 iff either: it's possible that K_1 and K_2 should be perceived together; or, there is some K_i such that it's possible that K_1 and K_i should be perceived together, and it's possible that K_i and K_2 should be perceived together.

Put in the simplest terms possible, the hylomorphic interpretation analyzes simultaneity in terms of joint perceptibility. Since kinetic cuts are the matter of nows while perception provides their form, the inclusion of both kinetic cuts and perception in the definiens guarantees that strict simultaneity is defined here as a genuinely temporal relation. This is important both for the immediate purpose and also because the notion of strict simultaneity will be employed in the definitions of derivative temporal relations.

The definition formalizes the requisite notion of joint perceptibility as a transitive relation in order to provide for the simultaneity of events that are not jointly perceptible by a single agent (on account of their separation from one another by vast space or impenetrable barriers). Simultaneity is commonly taken to be an equivalence relation: one that is reflexive (every now is simultaneous with itself), symmetric (pairs of nows are

[10] As I argued in section 7.2, the "when" specified at the beginning of this sentence is simply whenever the agent is in a waking state.

simultaneous with each other, if at all), and transitive (if now N_1 is simultaneous with now N_2, and N_2 is simultaneous with now N_3, then N_1 is simultaneous with N_3). The definition above likewise construes joint perceptibility as an equivalence relation,[11] and transitivity gives us what we want in the case of movements isolated from one another.[12] For example, a now included in the brushing of Smith's teeth in the USA will be jointly perceptible with a now included in the washing of Brown's dishes in the UK so long as there is a suitable series of intervening movements to function as the perceptible "links" between the two. As we shall see below, the movements of the celestial bodies are crucially involved in this and related matters.

The hylomorphic analysis of strict simultaneity is expressed in terms of pairs (or trios) of kinetic cuts. Having established that fundamental notion of simultaneity, we may now define an equally fundamental notion:

Kinetic world slice $=_{\mathrm{df}}$ a maximal union of strictly simultaneous kinetic cuts.

The matter of any now in the history of the cosmos is a particular kinetic world slice, which may be signified in set-theoretic terms as the union of ordered pairs whose elements are each an *F*-type *kinoumenon* considered as such (a !*kin*(*F*)), and a position on the *F*-bounded continuum at which the *kinoumenon* in question is actually located.[13] Indeed, *the* now is an individual kinetic world slice *qua* perceptible.[14]

11.2 PRIMARY TIME, BEING IN TIME

Aristotle gives an account of being in time in *Physics* IV.12 according to which to *be in* time is to *have a* time. More precisely, being in time requires having a "primary" time. Aristotle's notion of primary time is the

[11] Transitivity is explicitly built into the definition; reflexivity and symmetry get in by virtue of the fact that the definition places no restrictions on the identity of K_1 and K_2.

[12] Cf. White (1989), 212: "[A] time is a 'number' of an *equivalence class* of motions that have simultaneous 'prior' limits and simultaneous 'posterior' limits, that is, a time is an equivalence class of simultaneous motions considered as a (potential or actual) measure of motion."

[13] The specification of a kinetic world slice is certain to be far too long to be produced or comprehended by any human mind. Even limiting ourselves to clear *kinêta* (e.g. by disregarding the parts of organic compounds), the fact that any *kinêton* admits of multiple, non-equivalent descriptions has the consequence that every *kinêton* will appear in multiple elements of the union.

[14] This is not intended as an analysis of the property *being present*. I shall address the possibility of such an analysis on the hylomorphic interpretation in Chapter 12.

temporal analogue of his notion of place.[15] In *Physics* IV.4 he says, "Hence the place of a thing is the primary motionless boundary of what contains it" (212a20–1).[16] An object's place is the smallest surface of its surrounding environment beyond which no part of the object extends. We may define a movement's primary time by utilizing ">" as a kinetic order functor read as "is posterior to" when it joins kinetic world slice variables:

Primary time: the primary time of a movement *M* is the pair of nows having as their matter kinetic world slices S_1 and S_2 such that:

(a) S_1 includes no kinetic cut of M as an element
(b) S_2 includes a kinetic cut of M as an element
(c) $S_2 > S_1$
(d) there are no S_i such that:
  (i) $S_i > S_1$ and S_i fails to include a kinetic cut of *M* as an element
  (ii) $S_i > S_2$ and S_i includes a kinetic cut of *M* as an element.

Clause (d [i]) establishes kinetic world slice S_1 as the matter of the last moment before the existence of *M*'s *kinoumenon*, the last moment of rest.[17] Clause (d [ii]) establishes S_2 as the matter of the last moment of the *kinoumenon*'s existence, its arrival at the movement's *telos*. Intuitively, then, the primary time of a movement is determined by the pair of most proximate nows beyond which no part of the movement extends.

To be in time is to have a primary time, and the primary time of a movement "measures" the movement in the appropriate sense. This explains why Aristotle says that:

Further, a movement's being in time means that both it and its essence are measured by time (for it measures the movement and its essence together [*hama*], and this is what being in time means for it, that its essence should be measured). (221a4–7)[18]

The distinction between a movement and its essence might seem inscrutable outside of the hylomorphic interpretation. A movement is

[15] The analogy isn't perfect, since Aristotle explicitly rejects a hylomorphic analysis of place in *Phys.* IV.4 (212a3–5). The difference between place and primary time, as we shall see, is that the bounding surface definitive of place is intrinsically fully determinate, whereas its temporal analogue definitive of primary time is not.

[16] ὥστε τὸ τοῦ περιέχοντος πέρας ἀκίνητον πρῶτον, τοῦτ᾽ ἔστιν ὁ τόπος.

[17] Recall from section 5.4 that Aristotle denies that there is any first moment of a movement. Note also that clause (c) reflects my claim in section 5.4 that Aristotle's application of the notion of primary time in *Phys.* VI.5 to individual kinetic cuts is derivative.

[18] καὶ ἔστιν τῇ κινήσει τὸ ἐν χρόνῳ εἶναι τὸ μετρεῖσθαι τῷ χρόνῳ καὶ αὐτὴν καὶ τὸ εἶναι αὐτῆς (ἅμα γὰρ τὴν κίνησιν καὶ τὸ εἶναι τῆς κινήσεως μετρεῖ, καὶ τοῦτ᾽ ἔστιν αὐτῇ τὸ ἐν χρόνῳ εἶναι, τὸ μετρεῖσθαι αὐτῆς τὸ εἶναι).

individuated by a unique *kinoumenon*; the essence of a *kinoumenon* is the telic property that individuates the *kinoumenon* as such. *Kinoumena* are accidental compounds whose material component is a substance and whose form is a telic property.

11.3 THREE DERIVATIVE SIMULTANEITY RELATIONS

Let us use the fundamental simultaneity relations defined in section 11.1 above to derive three additional notions of simultaneity. First we may define a relation of weak simultaneity obtaining between entire movements:

Weak simultaneity: movement M_1 is weakly simultaneous with movement M_2 iff there is at least one kinetic world slice that includes a kinetic cut of M_1 and a kinetic cut of M_2 as two of its elements.[19]

This is the notion of partial temporal overlap. Intuitively, the idea is that two movements are weakly simultaneous iff they share at least one strictly simultaneous kinetic cut. Of course, because Aristotle rejects the idea that any movement may have a first moment of a movement, the condition specified in the definiens will, as a matter of nomological necessity, always be satisfied by a plurality of simultaneous kinetic cuts.

A stronger relation is that of exact temporal overlap:

Strong simultaneity: movement M_1 is strongly simultaneous with movement M_2 iff the kinetic world slices including kinetic cuts of M_1 as elements are exactly those kinetic world slices including kinetic cuts of M_2 as elements.

This is Aristotle's notion of "one and the same" time. Two events or movements occur in (or have) one and the same time just in case their limiting kinetic cuts are strictly simultaneous – that is, if they share precisely the same primary time. Any pair of movements that are strongly simultaneous are *ipso facto* weakly simultaneous, but the converse obviously does not hold.

Strong simultaneity provides the basis for analyzing Aristotle's notion of times that are "one in form":

Formal simultaneity: movement M_1 is formally simultaneous with movement M_2 iff either: M_1 and M_2 are strongly simultaneous; or, M_1 and M_2 are

[19] Note that neither this nor the following analysis requires that the M_i be numerically distinct. A consequence of this is that every movement is both weakly and strongly simultaneous with itself.

each strongly simultaneous with numerically distinct tokens of a type-repeatable motion M_K.[20]

Formal simultaneity captures Aristotle's view that time is a measure of motion in the robust sense of "measure" discussed in section 6.2. If we regard the emptying of a particular *klepshydra* as a type-repeatable motion, fairness in court proceedings may be effected by measuring the testimony given by litigants with distinct tokens of that motion type. To measure each litigant's speech in this way is to guarantee that each gets the same amount of time to speak.[21] Practices such as this suppose that distinct tokens of the type-repeatable motion are themselves formally simultaneous, which might be thought to raise the specter of skepticism. But as Aristotle argues in the *Physics* and elsewhere, there is a kind of motion fit to serve as the ultimate standard for all others.

11.4 CELESTIAL MOTION

Time is a number of motion. In *Physics* IV.14 Aristotle asks of what sort of motion time is the number, to which he responds that time is the number of any kind (*hopoiosoun*) of motion, not of some particular kind (*ou tinos*). But this liberal position is contradicted in the *Generation and Corruption*, where he maintains that "Time, therefore, is a number of some continuous movement – a number, therefore, of the circular movement [of the celestial bodies]" (337a24–6).[22] The tension between these two passages is relieved, if only somewhat, by Aristotle's subsequent concession in *Physics* IV.14 that "regular circular motion is above all else [*malista*] the measure, because the number of this is the most knowable [*gnôrimôtatos*]" (223b19–20).[23] Indeed, he says, the popular view that time just is the movement of the sphere is not altogether wrong, because "other movements are measured by this, and time is also measured by this movement" (223b21–3).[24]

[20] Note that a limiting case of formal simultaneity is when the M_i are themselves tokens of the type-repeatable motion M_K in question. Since every movement is strongly simultaneous with itself, any pair of such tokens are in fact (individually) strongly simultaneous with numerically distinct tokens of the type-repeatable motion and therefore are formally simultaneous with each other, are one in form.

[21] Cf. White (1989), 214.

[22] συνεχοῦς ἄρα τινὸς ἀριθμὸς ὁ χρόνος, τῆς κύκλῳ ἄρα, καθάπερ ἐν τοῖς ἐν ἀρχῇ λόγοις διωρίσθη.

[23] ἡ κυκλοφορία ἡ ὁμαλεῖς μέτρον μάλιστα, ὅτι ὁ ἀριθηὸς ὁ ταύτης γνωριμώτατος.

[24] διὸ καὶ δοκαῖ ὁ χρόνος εἶναι ἡ τῆς σφαίρας κίνησις, ὅτι ταύτῃ μετροῦνται αἱ ἄλλαι κινήσεις καὶ ὁ χρόνος ταύτῃ τῇ κινήσεσι.

I take Aristotle's claim that the number of celestial motion is most knowable (*gnôrimôtatos*) to be based partly upon the obvious fact that the heavenly bodies are visible from practically every location on the face of the earth. The ubiquity (and consequent familiarity) of celestial observations makes celestial movements an obvious perceptual "link" serving the requisite theoretical role in grounding simultaneity between spatially remote events. For instance, Coriscus notes that event *a* (which he witnesses in Athens) is strongly simultaneous with the movement of the stylus' shadow between two marks on a sundial; Callias notes that event *b* (which he witnesses in Thebes) is likewise strongly simultaneous with the corresponding movement of a local sundial's shadow.[25] Because the movements of the two sundials record a single movement of the sun, and because strong simultaneity is a transitive relation, events *a* and *b* are strongly simultaneous with one another. Only when one is out of view of the sun (say, in a courtroom) does one need to employ a chronometer that doesn't rely directly on the movement of the sun or some other heavenly body to keep time.

And yet this is only one sense of "knowable" as it functions within this context. A passage from the *Metaphysics* provides the second, and arguably more significant, sense. In Δ.6 Aristotle articulates several distinct meanings of "one," one of which bears directly on the issue under consideration:

> While in a sense we call anything one if it is a quantity and continuous, in a sense we do not unless it is a whole, i.e. unless it has one form. This is why the circle is of all lines most truly one, because it is whole, i.e. complete. What it is to be one is to be a principle of number; for the first measure is the principle, for that by which we first know [*prôtô gnôrizomen*] each genus is the first measure [*prôton metron*] of the genus; the one, then, is the principle [*archê*] of the knowable regarding each genus. But the one is not the same in all genera. For here it is a quartertone, and there it is the vowel or the consonant; and there is another unit of weight and another of movement. (1016b11–23)[26]

[25] Properly speaking, the movement of the stylus' shadow would need to be subject to longitudinal compensation for this illustration to be fully successful. In a similar vein, I ignore discrepancies resulting from differences in latitude.

[26] ἔτι δ' ἔστι μὲν ὡς ὁτιοῦν ἕν φαμεν εἶναι ἂν ᾖ ποσὸν καὶ συνεχές, ἔστι δ' ὡς οὔ, ἂν μή τι ὅλον ᾖ, τοῦτο δὲ ἂν μὴ τὸ εἶδος ἔχῃ ἕν· οἷον οὐκ ἄν φαῖμεν ὁμοίως ἓν ἰδόντες ὁπωσοῦν τὰ μέρη συγκείμενα τοῦ ὑποδήματος, ἐὰν μὴ διὰ τὴν συνέχειαν, ἀλλ' ἐὰν οὕτως ὥστε ὑπόδημα εἶναι καὶ εἶδός τι ἔχειν ἤδη ἕν· διὸ καὶ ἡ τοῦ κύκλου μάλιστα μία τῶν γραμμῶν, ὅτι ὅλη καὶ τέλειός ἐστιν. τὸ δὲ ἑνὶ εἶναι ἀρχῇ τινί ἐστιν ἀριθμοῦ εἶναι· τὸ γὰρ πρῶτον μέτρον ἀρχή, ᾧ γὰρ πρώτῳ γνωρίζομεν, τοῦτο πρῶτον μέτρον ἑκάστου γένους· ἀρχὴ οὖν τοῦ γνωστοῦ περὶ ἕκαστον τὸ ἕν. οὐ ταὐτὸ δὲ ἐν πᾶσι τοῖς γένεσι τὸ ἕν. ἔνθα μὲν γὰρ δίεσις ἔνθα δὲ τὸ φωνῆεν ἢ ἄφωνον· βάρους δὲ ἕτερον καὶ κινήσεως ἄλλο. Aristotle's casual shift from the language of number to the language of measure here presents a challenge for Coope's (2005), 87,

Aristotle doesn't specify what the unit of movement is, but his remark that "the circle is of all lines most truly one" provides excellent reason to regard circular motion as "first measure" of all other motion.

So it is the completeness of the circular motion of the celestial spheres that makes it the principle or source for measuring all other movements. But there is another feature of celestial motion that renders it uniquely fit to serve as the ultimate temporal standard: the matter of the heavenly bodies themselves. The sun, moon, and stars (and presumably the invisible rotating spheres in which they are embedded) are composed of a type of matter that is very different from the four elements found in the sublunary realm, a "quintessence" whose name, Aristotle tells us, derives from "always runs" (*aei thein*).[27] *Aether* is an exotic kind of stuff, being ungenerated, incorruptible, and pure.[28] That the matter of the heavenly bodies, and that each body itself, possesses such features is evident, Aristotle says, both to reason and to observation. All men (even barbarians) reserve the highest place in the cosmos for the immortal, divine beings,[29] and there is no place higher than the celestial spheres.[30] Moreover, no one in human history, he claims, has ever observed any change take place in the heavens other than its continuous circular motion.[31]

Therefore, Aristotle had good reason to embrace the view that the heavenly bodies are profoundly different from terrestrial bodies in that they are ungenerated and incorruptible, free from qualitative change of any kind. They also differ from their sublunar counterparts in terms of their respective natural motion. Each of the four familiar elements has a type of motion that is proper to its kind, which is determined by the natural *place* for each element – extremely upward for fire, extremely downward for earth, and intermediately for air and water. And so the natural motion for each of these elements is motion in a straight line either toward or away from the center of the earth until it reaches its natural place in the cosmos, its place of rest. Unlike the four terrestrial elements, *aether* has no place of rest, and its natural motion is circular.[32]

Since it is the very nature of the ungenerated and incorruptible *aether* to move in a perfect circle, and since the ungenerated heavenly bodies are

claim that Aristotle regards time essentially as a number of motion, but only accidentally as a measure of motion.

27 *Cael.* I.3, 270b23. 28 *Cael.* I.3, 270a12 ff., *passim*. 29 *Cael.* I.3, 270b5–10.

30 *Cael.* I.9, 279a12–18. In this chapter Aristotle also claims that the term for "duration" (αἰών) derives from the phrase "always existing" (αἰεὶ ὤν) (279a27).

31 *Cael.* I.3, 270b13–14.

32 *Cael.* I.2, 269a27–9b17. Note that Aristotle distinguishes the rotary motion of the celestial spheres from ordinary (terrestrial) circular motion at *Phys.* VIII.8, 262a15–16.

composed of nothing but *aether*, they forever have and always will move in a perfect circle, and by their very nature. The naturalness of this motion and the purity of the bodies that so move guarantees that the motion in question will be perfectly smooth and uniform throughout. This, then, is the basis for Aristotle's claim in *Physics* IV.14 that celestial movement is "regular" (*homaleis*), thereby providing suitable type-repeatable movements (designated by "M_K" in the definition of formal simultaneity in section II.3 above) to serve as the ultimate standard for temporally separated movements that are one in form.

This view, we should note, bears a strong affinity with the view endorsed by Plato in the *Timaeus*, according to which the numerically regular movements of the heavenly bodies ("movement-according-to-number") constitutes an image of eternity. Because the demiurge could not – on pain of contradiction – *create* something that is eternal, the very best he could do is to create something that mimics eternity by being the same as itself again and again in a never-ending process. In the *De Anima*, Aristotle clearly embraces this variety of surrogate for eternity in connection with animal and plant life:

> For the most natural of all functions for a living thing, if it is complete and not defective and does not come to be by chance, is to produce another thing of the same sort as itself ... in order to share as far as it can in the everlasting and divine ... These living things cannot share in the everlasting and divine by continuously existing, since no perishable thing can remain numerically one and the same; hence they share in it as far as they can, to different degrees, and what remains is not the [parent] itself, but something else of the same sort as [the parent] – something that is specifically, not numerically, one with [the parent].[33] (415a26–b8)

Despite this structural similarity of Aristotle's biological view and the image of eternity that Plato describes in the *Timaeus*, it would be a mistake to think that Aristotle's views concerning celestial motion are consistent with Plato's. For as we have already seen, Aristotle denies that the heavenly bodies are generated, and so on his view there was (to take just one example) no first circuit of the sun, whereas Plato is committed to the occurrence of just such an event. But beyond that difference, there is

[33] φυσικώτατον γὰρ τῶν ἔργων τοῖς ζῶσιν, ὅσα τέλεια καὶ μὴ πηρώματα, ἢ τὴν γένεσιν αὐτομάτην ἔχει, τὸ ποιῆσαι ἕτερον οἷον αὐτό ... ἵνα τοῦ ἀεὶ καὶ τοῦ θείου μετέχωσιν ᾗ δύνανται· ... ἐπεὶ οὖν κοινωμεῖν ἀδυνατεῖ τοῦ ἀεὶ καὶ τοῦ θείου τῇ συνεχείᾳ, διὰ τὸ μηδὲν ἐνδέχεσθαι τῶν φθαρτῶν ταὐτὸ καὶ ἓν ἀριθμῷ διαμένειν, ᾗ δύναται μετέχειν ἕκαστον, κοινωνεῖ ταύτῃ, τὸ μὲν μᾶλλον τὸ δ' ἧττον, καὶ διαμένει οὐκ αὐτὸ ἀλλ' οἷον αὐτό, ἀριθμῷ μὲν οὐκ ἕν, εἴδει δ' ἕν.

a more significant point to be drawn out in distinguishing the eternality of species versus the eternality of celestial motion. In the closing passages of the *Generation and Corruption*, Aristotle indicates that transits of the sun bear a rather stronger relation to one another than mere formal simultaneity:

> Thus, since the upper movement is cyclical, the sun moves in this determinate manner; and since the sun moves thus, the seasons in consequence come to be in a cycle, i.e. return upon themselves; and since they come to be cyclically, so in their turn do the things whose coming to be the seasons initiate ... Now it is evident that those things [that return upon themselves] whose substance – that which is undergoing the process – is imperishable, will be numerically the same; for the character of the process is determined by the character of that which undergoes it. Those things, on the other hand, whose substance is perishable (not imperishable) must return upon themselves specifically, not numerically. That is why when water comes to be from air and air from water, the air is the same specifically, not numerically.[34] (338b3–18)

This passage is challenging, because in it Aristotle appears to endorse the prima facie incredible view that there are not multiple, numerically distinct transits of the sun, that the sun has orbited the earth exactly once. This reading might be thought to gain support from a remark that Aristotle makes in the *Physics*: "as a movement can be one and the same again and again [*autên kai mian palin kai palin*], so too can time, e.g. a year or a spring or an autumn" (220b12–14).[35] But surely this really is too incredible to be the correct reading of the passage. What, then, is numerically unique, if not transits of the sun?

The answer is *the movement of the sun* (and other celestial bodies).[36] Aristotle's point in this passage is to distinguish biological reproduction and celestial motion as importantly different types of cyclical processes. Both processes are eternal, but they differ in two related ways. First, the respective subjects of the two process types differ fundamentally in character: plants and animals are materially corruptible (being composed of elements whose natural motions are in conflict), whereas the heavenly

[34] οἷον τῆς ἄνω φορᾶς οὔσης κύκλῳ ὁ ἥλιος ὡδί, ἐπεὶ δ᾽ οὕτως, αἱ ὧραι διὰ τοῦτο κύκλῳ γίνονται καὶ ἀνακάμπτουσιν, τούτων δ᾽ οὕτω γινομένων πάλιν τὰ ὑπὸ τούτων ... ὅσων μὲν οὖν ἄφθαρτος ἡ οὐσία ἡ κινουμένη, φανερὸν ὅτι καὶ ἀριθμῷ ταὐτὰ ἔσται (ἡ γὰρ κίνησις ἀκολουθεῖ τῷ κινουμένῳ), ὅσων δὲ μὴ ἀλλὰ φθαρτή, ἀνάγκη τῷ ἔδει, ἀριθμῷ δὲ μὴ ἀνακάμπτειν. διὸ ὕδωρ ἐξ ἀέρος καὶ ἀὴρ ἐξ ὕδατος εἴδει ὁ αὐτός, οὐκ ἀριθμῷ.

[35] ὡς ἐνδέχεται κίνησιν εἶναι τὴν αὐτὴν καὶ μίαν πάλιν καὶ πάλιν, οὕτω καὶ χρόνον, οἷον ἐναιυτὸν ἢ ἔαρ ἢ μετόπορον. Cf. *GA* II.1, 731b33–5.

[36] The sun itself is of course also numerically unique, but the *GC* passage is concerned centrally with the sun's movement.

bodies are not (being composed of a single, divine element whose natural motion is unitary and consistent).

But since "the character of the process is determined by the character of that which undergoes it," the former difference entails a difference in process. Biological reproduction is intrinsically serial, replete with beginnings, endings, and all forms of qualitative changes.

On the other hand, the movement of the sun (focusing again on just one example) is not intrinsically serial. Indeed, Aristotle believes that the sun has forever been undergoing a single, uninterrupted movement about the earth. The series of days with which we are intimately familiar are *parts* of this single movement of the sun. These parts have a peculiar status, being undetached from the whole of which they are parts, and being such that their magnitude and location within the whole are contingent upon particular facts about timekeeping agents. Time, as I have said previously, is precisely this determinability of motion, its susceptibility to being divided into undetached parts of arbitrary length. By adopting a convention to divide the sun's single movement into segments, *we* provide the missing conditions that are necessary for seriality. This is why I claimed that Aristotle does not regard the movement of the sun as being intrinsically serial.

CHAPTER 12

Temporal passage

Recall from the discussion in the Introduction that temporal passage is one of the three basic elements of the contemporary concept of time and consists in the apparent (or real) "flow" of time from past to future. A natural question to ask about Aristotle's temporal theory is whether he regards this "flow" as a genuine feature of reality. Richard Sorabji and Fred Miller both hold the view that Aristotle's account does incorporate temporal passage, but only equivocally and confusedly so. As Sorabji says:

> I am doubtful that he ever sensed the difference between static and flowing terminologies sufficiently clearly to envisage a solution [to the puzzles of IV.10] that turned on the distinction.[1]

He goes on to say, "without sharply distinguishing the flowing from the static, Aristotle none the less treats flowing conceptions as essential to the existence of time."[2] Sorabji maintains that the distinction between flowing and static conceptions of time was not made fully explicit until the Neoplatonist Iamblichus, some six and a half centuries after Aristotle had composed his own theory.[3]

Miller's view is consistent with Sorabji's. He claims that:

> This account [in IV.13] of the role of the now as a unifying and diverse principle of time is evidently quite different from the account presented earlier in ch. 11 at 220a5–21, though Aristotle seems unaware of the difference.[4]

[1] Sorabji (1983), 47. [2] Sorabji (1983), 50. [3] Sorabji (1983), 51.

[4] Miller (1974), 143. I must not neglect to acknowledge that I agree with much of what Miller has to say about Aristotle's account of time. Indeed, his brief description of the view on p. 136 sounds very much like my own interpretation in outline. Nonetheless, Miller is much less optimistic about the viability of Aristotle's reductive program (see esp. p. 147). Though he does not explicitly say so, Miller seems to agree with the judgment of Owen, Annas, and Corish that Aristotle's scheme is patently circular: "It is pretty clear where Aristotle has gone wrong in ch. 11: When we say that motion is always other and other, what we mean by this is that the motion consists in a subject possessing different accidents *at different instants*" (154). Perhaps Miller is correct in this claim, but Aristotle's account is not an attempt to analyze the meaning of statements about motion, or

I concede that Sorabji and Miller are right to claim that Aristotle does not explicitly distinguish the concepts of temporal extension and temporal passage. But it's also clear that their respective interpretations of these two chapters exaggerate the appearance of confusion, each being founded on a mistaken understanding of Aristotle's larger view concerning the nature of time. Aristotle simply doesn't *need* to raise the issue of temporal passage as an independent topic of discussion, because it is in a fairly straightforward sense built into the very nature of time. If I am able to show that Sorabji and Miller are wrong to claim that these passages present time as static, that they do in fact include a significant dynamic aspect, then Aristotle's view will have been defended against charges of equivocation and confusion. Supposing that this can be done, the remaining task will be to explain the nature of temporal passage within Aristotle's view of time. But first, the defensive project.

12.1 SORABJI'S "ANALOGIES"

Aristotle makes a remark early in his discussion of the now that appears to support the opinion that his view of time includes temporal passage: "And just as a movement is always other and other, so too is time" (219b9–10).[5] The phrase "always other and other" calls to mind something ever-changing in nature, and the notion of ever-changing time certainly suggests temporal "flow." Aristotle proceeds to address the question regarding the similarity or difference of the now (raised in one of the puzzles presented in IV.10, to be discussed in Chapter 13), and he answers in his typical fashion:

> The now is in a way the same, and in a way not the same. Insofar as it is other and other, it is different (that's what it is for it to be a now), but it is the same in respect of its substrate. (219b12–15)[6]

Sorabji balks at this manner of drawing the distinction, claiming that this is "almost the opposite" of the correct view:

> What needs to be said is that the instants are ever different, but the character of presentness is ever the same. Pseudo-Archytas comes closer to saying this. His formulation is that the now is *numerically* diverse, but remains the same *in form*.[7]

any other kind of statements. In addition to this general misunderstanding of Aristotle's project, some of Miller's remarks on the now and temporal passage are off the mark.

[5] καὶ ὥσπερ ἡ κίνησις αἰεὶ ἄλλη καὶ ἄλλη, καὶ ὁ χρόνος.

[6] τὸ δὲ νῦν ἔστι μὲν ὡς τὸ αὐτὸ, ἔστι δ' ὡς οὐ τὸ αὐτό· ᾗ μὲν γὰρ ἐν ἄλλῳ καὶ ἄλλῳ, ἕτερον (τοῦτο δ' ἦν αὐτῷ τὸ νῦν εἶναι), ὅ δέ ποτε ὄν ἐστι τὸ νῦν, τὸ αὐτό.

[7] Sorabji (1983), 48. Cf. Ross (1936), 68: "Aristotle would have done better, it seems, if instead of saying that what is a now is always (specifically) the same but to be now is different, he had said

Without further elaboration of what Pseudo-Archytas means by "same in form," it is difficult to know precisely what Sorabji's complaint and recommendation amount to. But we needn't press that point, because Sorabji further develops his objection by claiming that Aristotle "makes no less than four comparisons between the now and a moving body, but surprisingly never manages to articulate the sense in which the now is moving."[8] I maintain that: (i) Sorabji is quite wrong to read these remarks as nothing more than analogies; (ii) he is looking for something in Aristotle's account that Aristotle was right not to include; and (iii) the philosophically respectable version of what Sorabji wishes to find is in fact present in the passages in question.

I must postpone my discussion of (ii) to section 12.3, and I cannot completely address (iii) before having done so. Still, we should be able straightaway to establish (i) in full and to get a reasonable purchase on (iii). In order to do so, we need to examine each of Sorabji's "analogies" and the passages on which he bases them. Let us first consider his third "analogy," since doing so will put us on the path to seeing the error in the remaining three: "[T]ime and the now depend on each other for existence, as do motion and the moving body."[9] The passage he paraphrases here runs from 219b33 to 220a3:

> And so it's clear that if time didn't exist, the now wouldn't either, and that if the now didn't, time wouldn't. For just as the moving thing and the locomotion are together [*hama*], so too are the number of the moving thing [*tou pheromenou*] and the number of the locomotion.[10]

Aristotle claims in the first sentence that time and the now are dependent on one another for their existence. His explanation for this interdependence (*gar* – "for") takes the form of a pair of pairs whose members are related by being "together."

Now with respect to the first pair (the moving thing and the locomotion), we have seen in what sense the members are together.[11] Aristotle defines motion as the actuality of a *kinoumenon*, and so motion is logically posterior to (and therefore metaphysically dependent upon) *kinoumena* – no *kinoumenon*, no movement. As we have also seen, Aristotle adduces

that while the various things that are successively now are numerically different, their being now is specifically the same. He might have said truly that the flow of time consists in the inheritance by one moment from another of an identical character of nowness or presentness."

[8] Sorabji (1983), 49. [9] Sorabji (1983), 49.

[10] φανερὸν δὲ καὶ ὅτι εἴτε χρόνος μὴ εἴη, τὸ νῦν οὐκ ἂν εἴη, εἴτε τὸ νῦν μὴ εἴη, εἴτε τὸ νῦν μὴ εἴη, χρόνος οὐκ ἂν εἴη· ἅμα γὰρ ὥσπερ τὸ φερόμενον καὶ ἡ φορά, οὕτως καὶ ὁ ἀριθμὸς ὁ τοῦ φερομένου καὶ ὁ τῆς φορᾶς.

[11] See section 4.3. Recall that "*pheromenon*" names a species of *kinoumenon*.

independent reasons to reject the existence of instantaneous *kinoumena* so that the existence of a *kinoumenon* metaphysically necessitates the existence of a movement – no movement, no *kinoumenon*.[12] So the sense of "together" in which a movement and a *kinoumenon* are together is just the sense of mutual dependence that Aristotle believes to obtain between time and the now and is attempting to explain.

As we turn to the second pair in Sorabji's "analogy" (the number of the moving thing and the number of the locomotion), it is evident that the description "the number of the locomotion" is simply a reference to time that implements the theoretical terminology introduced in IV.11. Given the structure and aim of the explanation, this provides excellent reason to think that the expression "the number of the moving thing" is in this context Aristotle's way of referring to the now. But if this is correct, we ought to reject Sorabji's complaint that Aristotle is merely drawing a comparison between the interdependence of moving objects and movements on the one hand, the now and time on the other. Rather, having already established that time is an aspect of (and hence metaphysically and logically dependent upon) motion, Aristotle is claiming here that the now is an aspect of (and hence metaphysically and logically dependent upon) a *kinoumenon*: the now is "the number of the moving thing." Therefore, Sorabji's reading of this passage as an analogy is far too weak to be correct. The relation between time and motion that Aristotle invokes here is one of ontological dependence, not merely one of similarity. Time and the now depend on one another for their existence precisely because the same is true of movements and objects in motion. The significance of this is that Sorabji's claim that Aristotle fails "to bring out a sense in which the now is moving" is plainly false; indeed, it is a conceptual truth on Aristotle's view that the now involves movement, since its definition makes reference to *kinoumena*.

If the third of Sorabji's four "analogies" is unfounded, none of the remaining three fares any better. His first "analogy" is this:

> [A] moving object remains the same in respect of its *substratum* (it is still a stone, for example) even though it changes in its description or *logos* (it is now here, now there). The point of the analogy turns out to be not the *motion* of the stone, but its *sameness*.[13]

Sorabji accurately reports Aristotle's analysis of the relation between a *kinoumenon* and its material substrate, but as I showed in Chapter 4, the

[12] See section 5.4. [13] Sorabji (1983), 49.

material substrate of the now *is* a *kinoumenon* considered as such, not simply the *hupokeimenon* of the *kinoumenon*.[14] So here again, Sorabji's complaint that Aristotle doesn't characterize the now as involving motion is without basis.

The second "analogy" is this: "[W]e recognise before and after by means of the now, as we recognise before and after in motion by means of the moving body."[15] As we saw in Chapter 7, our ability to perceive the now is a critical aspect of Aristotle's view. However, he does not ever suggest that the now is *instrumental* in our recognition of the before and after (whether in motion or in time), as Sorabji's use of the phrase "by means of" suggests. What Aristotle says is this:

> But the now follows the thing in motion, just as time follows motion (for we come to know the before and after in motion by means of the moving thing, and the now is the before and after *qua* numerable). (219b22–5)[16]

As we have seen, kinetic cuts are states of *kinoumena*, which provide the basis for the distinction between the before and after in motion and, consequently, in time. Thus, again, the moving body is partly constitutive of the before and after, both in motion and in time. So here as well, Sorabji's reading of the passage as an analogy is too weak, and his complaint is mistaken.

Sorabji's final "analogy" also misrepresents Aristotle's view: "[T]he now both joins together and divides past and future, which is what the moving body does in relation to motion."[17] We have examined Aristotle's own words once before, and they are these:

> Indeed, time is both made continuous by the now and divided by it. And this follows the locomotion and the thing in motion. For both the motion and the locomotion are made one by the moving thing, because it is one (not one in substrate – for it might leave a gap in the movement – but rather one in account), and this determines the before and after in the movement. (220a4–9)[18]

[14] More properly, the material substrate of a now *includes* a *kinoumenon*, since it is a kinetic cut that constitutes the matter of a now. In a footnote to this "analogy," Sorabji denies (without argument) that the substrate of the now could be a *kinoumenon*. One might suppose that he has here conflated the distinction between the substrate of the now (which is a *kinoumenon*) and that of the *kinoumenon* (which is typically an *ousia*, a substance). His fourth "analogy" also blurs the distinction.

[15] Sorabji (1983), 49.

[16] τῷ δὲ φερομένῳ ἀκολουθεῖ τὸ νῦν, ὥσπερ ὁ χρόνος τῇ κινήσει (τῷ γὰρ φερομένῳ γνωρίζομεν τὸ πρότερον καὶ ὕστερον ἐν κινήσει, ᾗ δ' ἀριθμητὸν τὸ πρότερον καὶ ὕστερον, τὸ νῦν ἔστιν).

[17] Sorabji (1983), 49.

[18] καὶ συνεχής τε δὴ ὁ χρόνος τῷ νῦν, καὶ διήρηται κατὰ τὸ νῦν· ἀκολουθεῖ γὰρ καὶ τοῦτο τῇ φορᾷ καὶ τῷ φερομένῳ. καὶ γὰρ ἡ κίνησις καὶ ἡ φορὰ μία τῷ φερομένῳ, ὅτι ἕν (καὶ οὐχ ὅ

Aristotle makes it very plain in this passage that movements are unified and individuated by *kinoumena* considered as such, not simply by unique substances.

So Sorabji's complaint that Aristotle fails to provide any sense in which the now is moving is without basis. While Aristotle does not claim that the now moves (as Sorabji recommends), he is in fact committed to the view that the now crucially involves motion insofar as its material component includes a moving object considered as such.

12.2 MILLER'S CHALLENGE

"It is a commonplace that Aristotle treats time as something inherently dynamic," Miller says, conceding that he does indeed so treat it in IV.11.[19] In this respect, he clearly has not fallen prey to Sorabji's misreading. However, Miller claims that Aristotle abandons his dynamic view of time in IV.13, characterizing time as something static, unaware that he altered his view. The passage underlying Miller's worry is found at the opening of chapter 13:

> The now is the connection of time, just as was said, for it connects time that has passed with time yet to be; and it is a limit of time, for it is a beginning of the one and the end of the other. But this isn't as obvious as it is with the point, which stays put. It divides potentially, and as such, the now is always different, but insofar as it binds together, it is always the same, just as it is in the case of mathematical lines (for it's not the case that a point is always the same in thought – dividing, it is always different; but insofar as it is one, it is the same in every way). (222a10–17)[20]

This is a difficult passage, but the discussion of division and unification clearly makes contact with the passage in IV.11 that formed the basis of Sorabji's fourth "analogy." In this passage, however, Aristotle seems to be making a different point from the one borne by his remarks in chapter 11. Miller's take on the situation is this:

ποτε ὄν – καὶ γὰρ ἂν διαλίποι – ἀλλὰ τῷ λόγῳ)· καὶ ὁρίζει δὲ τὴν πρότερον καὶ ὕστερον κίνησιν τοῦτο.

[19] Miller (1974), 147.

[20] Τὸ δὲ νῦν ἐστιν συνέχεια χρόνου, ὥσπερ ἐλέχθη· συνέχει γὰρ τὸν χρόνον τὸν παρεληλυθότα καὶ ἐσόμενον, καὶ πέρας χρόνου ἐστίν· ἔστι γὰρ τοῦ μὲν ἀρχή, τοῦ δὲ τελευτή. ἀλλὰ τοῦτ᾽ οὐχ ὥσπερ ἐπὶ τῆς στιγμῆς μενούσης φανερόν. διαιρεῖ δὲ δυνάμει. καὶ ᾗ μὲν τοιοῦτο, αἰεὶ ἕτερον τὸ νῦν, ᾗ δὲ συνδεῖ, αἰεὶ τὸ αὐτό, ὥσπερ ἐπὶ τῶν μαθηματικῶν γραμμῶν (οὐ γὰρ ἡ αὐτὴ αἰεὶ στιγμὴ τῇ νοέσει· διαιρούντων γὰρ ἄλλη καὶ ἄλλη· ᾗ δὲ μία, ἡ αὐτὴ πάντῃ).

> [A] fundamental feature of the ch. 11 account has so far been ignored: the assumption that time itself is in a process of becoming. To capture what Aristotle regards as the dynamic character of time, it is necessary to think of the analogue of the now, not as a geometrical point stationed somewhere along line *A*, but as a point that is *moving* along the line and occupying [distinct] points.[21]

He concludes that the passage from IV.11

> is in sharp contrast with the fixed-point analogy of ch. 13, which is thought to be instructive precisely because the point is fixed. In the case of the fixed-point analogy, the distinction ... appears to be a distinction between type and tokens; but in the case of the moving-point analogy, it appears to be a distinction between a substrate and its accidents.[22]

I believe that Miller is right both in thinking that the remark in IV.13 *is* an analogy (though like Sorabji, he also treats the passage in IV.11 as an analogy), and in claiming that Aristotle is making different points in the two passages. But Miller is wrong to claim that the passage in IV.13 portrays time as static. Whereas Sorabji's reading of Aristotle on this issue is rather too weak, Miller's reading of the analogy between the now and the mathematical point is far too strong. Aristotle certainly is not claiming that the now is like a mathematical point in every respect; indeed, he explicitly identifies one important disanalogy when he says that the point stays put, implying that the same is in some sense not true of the now. Miller correctly points out that the analogy is thought to be instructive because of the stability of geometrical lines and points, since Aristotle's principal aim in this passage is to explain how the process of conceptually dividing movements is similar to the process of bisecting lines, a process we examined in detail in section 6.2. But another aspect of its instructiveness lies in the particular difference between geometric points and nows that he mentions: the point stays put. So we are not required to attribute a static view of time to Aristotle simply for his having drawn the analogy.

In the IV.11 passage, Aristotle claims that the now is different insofar as the kinetic world slice answering to the designation "now" is always distinct. In the IV.13 passage, he is claiming that even a unique now can be thought of as a plurality in just the way that a mathematical point can – that is, by admitting of distinct, non-equivalent descriptions ("end of the past/left-hand line-segment," "beginning of the future/right-hand line-segment").[23] This is the significance of his claim that the mathematical

[21] Miller (1974), 145. [22] Miller (1974), 146.

[23] It is interesting to note that Aristotle conceives the timeline vertically, with the past occupying the upward direction, future downward – see *GC* II.11, esp. 338a9 ff.

point is "not … always the same in thought," an idea reminiscent of his discussion in Book III of the *De Anima* concerning the conceptual difficulty involved in the common sense-faculty's being a unity-in-difference:

> The fact is that just as what some thinkers describe as a point is, as being both one and two, in this sense divisible, so too insofar as the judging faculty is indivisible, it is one and instantaneous in action; but insofar as it is divisible, it uses the same symbol twice together. (427a9–13)[24]

Miller acknowledges that this is the import of the discussion at the opening of IV.13,[25] but instead of reading Aristotle as making the rather modest claim I have attributed to him – namely, that there are two senses in which the now is different – Miller seems intent on reading the passage in a way that is inconsistent with the remarks made in IV.11. Not only is such a reading not required by the text, it seems to me untenable.

Miller finds another problem in Aristotle's characterization of the now as the divisor of time:

> Since time is a continuum, it is perhaps acceptable to think of the now as a link [*sunechei*] between the past and the future. It is difficult, however, to think of the now as a genuine point of division. A four-dimensional spatiotemporal worm cannot be chopped literally into temporally separate segments in the way that an ordinary earthworm can be cut up into spatially separate parts. Of course Aristotle is only speaking of *potential* division [*dunamei*, *kata dunamin* – a14, 18], but this is not a clear notion in a case in which actual division is impossible.[26]

Here Miller seems to equate acts of division with acts of sundering. It is of course not required that an item be rendered asunder in order for it to have been divided; were this so, arithmetic division would be impossible. Rather, all that needs to be done in order to accomplish conceptual division (as opposed to physical sundering) of some continuous item is to treat a particular point – the point of division – as fulfilling the dual role mentioned above.[27] When Aristotle says in the passage above that the now divides potentially, all he is saying is that any now *may* be treated as fulfilling the dual role required for conceptual division, though it need not be. We have already examined this issue in some detail in terms of the distinction between determinability and determinacy, between time

[24] ἀλλ' ὥσπερ ἣν καλοῦσί τινες στιγμήν, ᾗ μία καὶ δύο, ταύτῃ καὶ διαιρετή. ᾗ μὲν οὖν ἀδιαίρετον, ἓν τὸ κρῖνόν ἐστι καὶ ἅμα, ᾗ δὲ διαιρετὸν ὑπάρχει, δὶς τῷ αὐτῷ χρῆται σημείῳ ἅμα.

[25] Miller, (1974) 142, 144.

[26] Miller (1974), 141–2. I have replaced the Greek terms in the original with transliterations.

[27] Cf. 220a12.

and times, and between measuring and numbering. So here, too, Miller's concern is unfounded.

Let this complete our critique of the view that Aristotle's treatment of time is not consistent with respect to the issue of temporal passage. Neither Sorabji nor Miller provides compelling reason to think that Aristotle's view is inconsistent or confused. Our remaining task is to explain just what Aristotle takes temporal passage to consist in.

12.3 ARISTOTELIAN TEMPORAL PASSAGE

Recall that Sorabji laments Aristotle's failure to "bring out a sense in which the now is moving." I claimed that Sorabji is demanding something of Aristotle that is illegitimate. That this is so is brought out very sharply by an argument for the impossibility of temporal passage made by D. C. Williams. After showing how Aristotle's view evades this type of objection, I shall consider a distinct kind of objection against the possibility of temporal passage made by Michael Dummett. The ability of Aristotle's view to deflect such objections is an indication of its conceptual sophistication and is something generally unappreciated by commentators.

12.3.1 Overcoming Williams-style objections

Temporal passage was characterized in Chapter 1 in terms of McTaggart's *A* series, which he employs as a formal representation of the intuitive view that time flows past us, or that we are in some way moving into the future along with the present moment. McTaggart argued that there is a contradiction inherent in the *A* series, and other philosophers have approved his case. D. C. Williams, in particular, makes no bones in renouncing those philosophers who are committed to the reality of the *A* series:

> [The friend of the *A* series is committed to a] proposition which is never more than vaguely expressed: that over and above the sheer spread of events, with their several qualities, along the time axis, which is analogous enough to the spread of space, there is something extra, something active and dynamic, which is often and perhaps best described as "passage." This something extra I think is a myth: not one of those myths which foreshadow a difficult truth in a metaphorical way, but one which is fundamentally false, deceiving us about the facts, and blocking our understanding of them.[28]

[28] Williams (1951), 460.

To see that the theory of temporal passage is false, Williams maintains, one need only consider the theory's central claim that the now moves.[29] It is widely held (Aristotle notwithstanding) that motion and change cannot take place outside of time: material objects move and undergo qualitative change in time. But if the now moves (or, variously, if instants undergo change with respect to their possession of the temporal properties *being future*, *being present*, and *being past*), it seems necessary to posit a "super-time" in which the now moves (or instantiates the temporal properties); material objects move in time, so the now must move in time of a higher order. This gives rise to a dilemma for the advocate of temporal passage: is this "supertime" a version of McTaggart's *A* series or his *B* series? If the former, then the friend of passage has set himself squarely on the path of a vicious infinite regress. If the latter, then the objector will demand an explanation for the disparity; if the *B* series is good enough for "super-time," why is it not good enough for ordinary time?

Remarkably, Norman Kretzmann has sketched an interpretation of Aristotle's theory of time that places him in the same camp as Williams and McTaggart. He argues, in effect, that there are no proper Aristotelian solutions to the puzzles Aristotle presents in IV.10. On the contrary, he says, they "make a very important contribution to the understanding of Aristotle's own theory of time."[30] Kretzmann says, "if I am right about the aim of the puzzles, the conclusions will constitute embarrassments for the familiar view that time is essentially, really *passing*."[31] What they show, then, is that Aristotle "takes time to be *essentially* the permanent ordering of events and that he considers the passage of time to be an attribute of, or the appearance of, that linear sequence."[32] Kretzmann, therefore, regards Aristotle as a "*B* theorist."

As my remarks in the preceding sections should make evident, I believe that Kretzmann's view cannot be correct, given what Aristotle says in the chapters that follow his presentation of the puzzles. Aristotle is clearly committed to a dynamic conception of time. But if I am correct about this, it seems that Aristotle is obliged to offer some kind of response to Williams' challenge. Williams elaborates on the incoherence of temporal passage as follows:

[29] Williams (1951), 463–4. This is essentially the same argument against passage that Broad (1967) considers – and subsequently rejects, calling it a "howler" – as a variation on McTaggart's argument for the irreality of time.

[30] Kretzmann (1976), 91. A detailed treatment of the puzzles will be provided in Chapter 13.

[31] Kretzmann (1976), 92.

[32] Kretzmann (1976), 107. Cf. 108: "I maintain, in place of [temporal passage] he holds that the essence of time is the fixed temporal order" of the *B* series.

The obvious and notorious fault of the idea, as we have now localized it, is this. Motion is already defined and explained in the dimensional manifold as consisting of the presence of the same individual in different places at different times … [However,] time "flows" only in the sense in which a line flows or a landscape "recedes into the west." That is, it is an ordered extension.[33]

The now cannot move, so the argument goes, because motion already involves time. The premise of this argument, it will be recognized, is the one implicitly employed by commentators who cannot see any way for Aristotle's definition of time to avoid circularity. But it was established in Chapter 4 that Aristotle denies the premise, and in so doing the possibility remains open for him to escape both the general charge of circularity as well as the absurdities inherent in the standard version of temporal passage characterized in terms of a moving now.

On Aristotle's view (contra Williams' characterization), motion is not defined in terms of time; rather, time is defined in terms of motion. And so just as time itself is a feature of motion, temporal passage must also be a feature of "kinetic passage." But of course "kinetic passage" is nothing other than *kinêsis*, garden-variety motion. Consequently, temporal passage enters Aristotle's theory by way of *kinoumena*, and the only sense in which the now moves is precisely the sense in which ordinary substances move. This is (as far as I can tell) all one could reasonably demand that Aristotle say by way of committing himself to the view that the now moves. The now is not a self-sufficient entity existing independently of objects in motion, not a peculiar sort of object whose motion needs to be accounted for by means of postulating higher-order temporal continua. Rather, the now is the natural result of percipient agents interacting with a world in motion. Hence, Aristotle is not claiming in the argument at 219b33–220a3 that temporal passage is a species of motion; on the contrary, he is claiming temporal passage is just "kinetic passage" insofar as it is perceptible. To notice that the now is different is just to notice that some object or other has undergone some kind of change.

Perhaps this line seems too easy. Miller, for example, surmises that although Aristotle is able to evade the Williams-style objection,[34] he does not do so without incurring an unacceptable cost:

However, although Aristotle may have taken out an insurance policy against this line of criticism … a Rylean worry suggests itself. Where one ordinarily thinks of a single event taking place, Aristotle has two concurrent events. Simultaneous with the process that consists in Coriscus walking from the Lyceum to the

[33] Williams (1951), 462–3. [34] Miller (1974), 153–4.

Agora runs another process that consists in the substrate now taking on various attributes. The treatment of time is supposed to function as part of a general explanation of change. But one does not make a process any more intelligible by postulating a second, even more mysterious, process that requires the same sort of explanation as the first. Here there is, clearly, "one process, not two."[35]

I cannot agree with Miller that Aristotle intends that his theory of time serve any explanatory function with respect to this kinetic theory; nothing is said in the presentation of either theory to suggest that he harbored such hopes. But aside from that fact, we must ask what Miller takes Aristotle's "substrate now" to be. Unfortunately, he does not give us a clear answer to this question. In the course of criticizing W. Wieland's interpretation, Miller denies that the now can be identical with the moving thing,[36] but he doesn't consider anything more specific than something that is an "analogue" of "a continuant, the body undergoing motion."[37] It is no wonder that Miller finds temporal passage to be wholly mysterious within Aristotle's scheme, his conception of the now's substrate is so vague.

Still, I agree with his opinion that within Coriscus' walking from the Lyceum to the Agora and the substrate now's taking on various attributes there is "one process, not two." What Miller fails to recognize is that Aristotle, too, shares this opinion. On the hylomorphic interpretation of Aristotle's view, a now is a kinetic cut (a state of a *kinoumenon*) insofar as it is perceptually individuable. An *F*-type movement consists, I have argued, in the actuality of an *F*-bound *kinoumenon* (a !*kin*(F)). But this selfsame *kinoumenon* is also the "engine" of temporal passage, insofar as its varied states are candidate objects of acts of awareness. So there is indeed a single process involved in Coriscus' walking from the Lyceum to the Agora and the substrate now's taking on various attributes, a process that Aristotle characterizes in a difficult but important passage from which Sorabji draws one of his "analogies":

And in these [the before and after in motion], too, the substrate of the now is the same (for it is *what is* before and after in motion), but it is different in being (for the now is the before and after *qua* numerable). And this is what is most knowable, for motion is known by means of the thing in motion, and locomotion

[35] Miller (1974), 154. Miller lists two other lines of criticism, the first of which mistakenly attributes to Aristotle a theory of meaning that is objectionably circular (see note 4 above), the second of which is a restatement of his claim that Aristotle abandons the dynamic characterization of time in IV.11 in favor of a static characterization in IV.13. I have already shown that neither of these two objections is compelling.

[36] Miller (1974), n. 10, and 148–9. [37] Miller (1974), 153.

by means of the thing in locomotion. For the thing in motion is a this, but the movement is not. And so the now is in one way always the same, and in another not the same, for this is also the case with the thing in motion. (219b26–33)[38]

A *kinoumenon* is *what is* before and after in a movement, and its being before and its being after consists in its possessing different properties (whether local, qualitative, or otherwise). This is what Aristotle means when he says that it is "different in being." Such states of *kinoumena* are kinetic cuts, and these just are nows insofar as they are individuable by means of perception. Thus, distinct nows are different in precisely the same way that distinct kinetic cuts are different: the *kinoumenon* is located at diverse spatial positions. The sense in which time passes, then, is exactly this: things move, and we are aware of their movements.

12.3.2 Overcoming Dummett-style objections

Michael Dummett characterizes his own argument as an alternative formulation of what he takes to be the fundamental idea underlying McTaggart's argument: "of anything that is real, there must be a complete – that is, observer-independent – description."[39] Since any adequate description of events in time requires the use of token-reflexive terms (such as "now"), as he claims, there is no observer-independent description of time. Thus, time is unreal.

One might well have serious concerns about the major premise of Dummett's argument. Thomas Nagel, for example, famously denies that there is a strict connection between reality and objectivity, maintaining instead that any complete conception of the world not only may, but *must* include at least some token-reflexives.[40] But I should like to set aside this thorny issue and attempt to provide exactly what Dummett (and, by extension, McTaggart) claims does not exist: a description of events as being in time without the use of token-reflexives.

Armed with an Aristotelian account of time, it is not difficult to see our way to such a description: in world *w*, *kinoumenon k* exists, and percipient

[38] ὥστε καὶ ἐν τούτοις ὃ μέν ποτε ὂν νῦν ἐστι, τὸ αὐτό (τὸ πρότερον γὰρ καὶ ὕστερόν ἐστι τὸ ἐν κινήσει), τὸ δ' εἶναι ἕτερον (ᾗ ἀριθμητὸν γὰρ τὸ πρότερον καὶ ὕστερον, τὸ νῦν ἔστιν). καὶ γνώριμον δὲ μάλιστα τοῦτ' ἔστιν· καὶ γὰρ ἡ κίνησις διὰ το κινούμενον καὶ ἡ φορὰ διὰ τὸ φερόμενον· τόδε γάρ τι τὸ φερόμενον, ἡ δὲ κίνησις οὔ. ἔστι μὲν οὖν ὡς τὸ αὐτὸ τὸ νῦν αἰεί, ἔστι δ' ὡς οὐ τὸ αὐτό· καὶ γὰρ τὸ φερόμενον.

[39] Dummett (1978), 356. [40] Nagel (1986), 4, 57.

agent *a* is capable of perceiving *k*. In *w*, *k* exists in time in the appropriate sense.[41]

Now, Dummett clearly would find this description inadequate, and for two reasons. First, he claims that any such static description or model cannot represent temporal order except by convention:

> [T]here is an element of convention in the … representation: we lay it down that the axes are to be chosen in a certain way, that such-and-such an axis represents time, and that such-and-such a direction along this axis represents the direction earlier-to-later; these conventions cannot be shown in the model.[42]

As we saw in Chapter 4, there is no such element of convention in the Aristotelian model. For the proper description of *k* as a *kinoumenon* will include *k*'s *telos*, which establishes an objective kinetic order that is in fact contained in the model itself. And since *k* is by hypothesis perceptible (by *a*), that objective kinetic order establishes an objective temporal order in *w*.

Dummett's second complaint alleges that such a description is simply incomplete:

> [W]e are sometimes inclined to suppose that what we observe at any one time is a three-dimensional segment of a static four-dimensional physical reality; but as we travel through the four-dimensional structure we observe different three-dimensional segments at different times. But of course the fourth dimension can no more be identified with time than the road down which someone travels can be identified with the time that passes as he travels down it. If our hypothetical observer observes only the four-dimensional configuration without observing our movement – the movement of our consciousness – through it, like someone observing the road but blind to the traveler, he does not see all that happens.[43]

We ought first to note that Dummett's desire that our consciousness be made to move seems unsettlingly similar to the demand that the present moment move, the very idea that Williams roundly (and rightly) ridicules. As we saw in the previous section, whatever degree of success that objection has against standard *A*-theoretic conceptions of time, it does not touch Aristotle's view.

Dummett's talk of dimensions is foreign to Aristotle's way of thinking. But if we were to adapt his view to it, it is clear that the fourth dimension consists of motion and change, not simply an array of events whose presence and location in the sequence is arbitrary so far as the dimension itself

[41] I believe this description is adequate to the task, but a more sophisticated description is available, one that would directly engage the account of primary time and being in time that I provided in section 11.3 above.

[42] Dummett (1978), 355. [43] Dummett (1978), 355.

is concerned. So when Dummett complains that the "movement of our consciousness" cannot be observed by inspecting the four-dimensional configuration, he is mistaken. After all, both motion and consciousness of that motion are *stipulated* in the description, and I can think of no argument for the incoherence of that description that doesn't simply beg the question against Aristotle's account of the nature of time. Relatedly, Dummett might be right when he says that it's a mistake to identify the road down which someone travels with the time it takes him to do so, but as Aristotle's account shows, some relation short of identity can serve well to connect temporal intervals with spatial intervals in a substantive and interesting way.

The conclusion of such considerations is striking: a theory about the nature of time constructed two and a half millennia ago is capable of weathering objections that are regarded by some philosophers as absolutely devastating to more contemporary theories of time. This result suggests that *A* theorists might do well to look to Aristotle's theory for inspiration in adapting their own best current theories. I shall speculate very briefly on this idea in the concluding chapter of this book. But for now we must address the question, by virtue of what fact is the present *present*?

12.4 THE ABSOLUTE PRESENT

The short answer to our question is that there is no further fact. While the hylomorphic interpretation yields fruit in connection with temporal extension (motion's determinability), simultaneity (joint perceptibility), and temporal passage (kinetic passage), it cannot provide any deep analysis of the absolute present. To be sure, the absolute present consists materially in a kinetic world slice, and formally in that kinetic world slice's individuability by means of perception. But this is true of any moment in history. Aristotle apparently takes it for granted that there is a way that the world is now and that there simply is no further fact to account for its being that way now.

It's important to understand precisely what is meant by that claim. There is no suggestion that the determinate way the world is is a "brute fact," on Aristotle's view. Quite to the contrary, Aristotle has a robust notion of causation capable of providing various and sundry explanations for why the world is the way it is at any given time. What his fourfold account of causation cannot provide – whether through the hylomorphic interpretation or otherwise – is any satisfactory explanation for why *this* kinetic world slice is absolutely present, as opposed to some other.

Nor is there any suggestion that Aristotle's unwillingness or inability to analyze presentness threatens his commitment to the reality of temporal passage. One might well be content to describe temporal passage within his theory as "the continuous projection of a property across events," but to do so requires one to acknowledge that presentness is a primitive property. That way of putting the point is misleading, though, since Aristotle wishes to account for temporal properties and facts in terms of more basic kinetic and perceptual properties and facts. So a more fitting description would capitalize on the world's uniqueness and determinacy with respect to its actual properties. "Actuality" is a primitive predicate that designates (among other things) a unique kinetic world slice. It should come as no great surprise to us that Aristotle makes no attempt to analyze actuality, since it figures as an analysans in so many of his analyses.

Indeed, actuality's resistance to analysis is analogous to the unprovability of the law of non-contradiction: axiomaticity is the conceptual/linguistic analogue of metaphysical primitiveness.[44] So even while Aristotle has a hylomorphic analysis of temporal passage, with respect to the status of the absolute present itself, we must say that he is an anti-reductivist *A* theorist: presentness is a genuine but unanalyzable feature of the structure of reality.

12.5 TIME'S CORROSIVE EFFECTS

Characterizing Aristotle's temporal theory in the reductive terms of hylomorphism might be thought to raise difficulties in accounting for his remarks about time's role in corruption and decay. For how could time cause anything to occur if it really is nothing more than motion's susceptibility to being divided into undetached parts of arbitrary length?

That Aristotle regards it as at least qualifiedly legitimate to speak of time as playing such a causal role is evident in *Physics* IV.12, where he says:

> Something, then, will be affected by time, just as we are accustomed to say that time wastes things away, and that all things grow old through time, and that people forget owing to the lapse of time, but we do not say the same of getting to know or of becoming young or beautiful. For by its nature time is rather the cause of decay, since it is the number of change, and change removes what is. (221a30–b2)[45]

[44] Cf. *Met.* Γ.4, where Aristotle maintains that the only arguments that can be given in support of LNC are *ad hominem*.

[45] καὶ πάσχει δή τι ὑπὸ τοῦ χρόνου, καθάπερ καὶ λέγειν εἰώθαμεν ὅτι κατατήκει ὁ χρόνος, καὶ γηράσκει πάνθ' ὑπὸ τοῦ χρόνου, καὶ ἐπιλανθάνεται διὰ τὸν χρόνον, ἀλλ' οὐ μεμάθηκεν,

Aristotle's invocation of his definition of time in the final sentence suggests that he endorses the ways of speaking described in this passage. Despite this apparent endorsement, he recants in the following chapter, where he indicates that, strictly and philosophically speaking, time as such is not the agent of decay:

> And for the most part this is what we customarily mean by a thing's being destroyed by time. Still, time doesn't even do this; rather, this kind of change just happens to occur in time. (222b24–7)[46]

This deflationary treatment of time's corrosive effects would indeed be rather disappointing if it amounts to nothing more than the claim that time destroys things only in the sense that destruction happens to occur in time.

Fortunately, the account has a bit more content than just that. To see what else is involved, we must look to the *Generation and Corruption*. As it turns out, the qualified legitimacy of talk about time's destructive powers is closely related to the qualified legitimacy of thinking that time is just the motion of the heavenly spheres (223b21–3), which as we saw in section 11.4 is not strictly correct, but also not so far from the truth.

While the circular motion of the celestial bodies cannot possibly be responsible for the decay of terrestrial bodies, because it is uniform in character, there is another aspect of celestial motion that can be – namely, the sun's movement along the ecliptic:

> This explains why it is not the primary motion that causes coming-to-be and passing away, but the motion along the inclined circle; for this motion not only possesses the necessary continuity, but includes a duality of motions as well ... Now the continuity of this motion is caused by the motion of the whole [cosmos]; but the approaching and retreating of the moving body are caused by the inclination ... And there are facts of observation in manifest agreement with our theories. Thus we see that coming-to-be occurs as the sun approaches and decay as it retreats; and we see that the two processes occupy equal times. (336a32–b18)[47]

οὐδὲ νέον γέγονεν οὐδὲ καλόν· φθορᾶς γὰρ αἴτιος καθ' ἑαυτὸν μᾶλλον ὁ χρόνος· ἀριθμὸς γὰρ κινήσεως, ἡ δὲ κίνησις ἐξίστησιν τὸ ὑπάρχον·

[46] καὶ ταύτην μάλιστα λέγειν εἰώθαμεν ὑπὸ τοῦ χρόνου φθοράν. οὐ μὴν ἀλλ' οὐδὲ ταύτην ὁ χρόνος ποιεῖ, ἀλλὰ συμβαίνει ἐν χρόνῳ γίγνεσθαι καὶ ταύτην τὴν μεταβολήν.

[47] Διὸ καὶ οὐχ ἡ πρώτη φορὰ αἰτία ἐστὶ γενέσεως καὶ φθορᾶς, ἀλλ' ἡ κατὰ τὸν λοξὸν κύκλον· ἐν ταύτῃ γὰρ καὶ τὸ συνεχές ἐστι καὶ τὸ κινεῖσθαι δύο κινήσεις· ... τῆς μὲν οὖν συνεχείας ἡ τοῦ ὅλου φορὰ αἰτία, τοῦ δὲ προσιέναι καὶ ἀπιέναι ἡ ἔγκλισις· ... Φαίνεται δὲ καὶ τὰ κατὰ τὴν αἴσθησιν ὁμολογούμενα τοῖς παρ' ἡμῶν λόγοις· ὁρῶμεν γὰρ ὅτι προσιόντος μὲν τοῦ ἡλίου γένεσίς ἐστιν, ἀπιόντος δὲ φθίσις, καὶ ἐν ἴσῳ χρόνῳ ἑκάτερον· Aristotle's inspiration here might once again be the *Ti*. At 35a ff. Plato characterizes circular celestial motion as "the Same" (τὸ αὐτό) and motion along the ecliptic as "the Different" (τὸ ἕτερον).

Here the generation and corruption Aristotle has in mind seem to be the seasonal growth and decay (or dormancy) of plants. The sun's heat in spring and summer serves as a catalyst for elemental combination, by virtue of which organisms come to be and grow. When the sun recedes in autumn and winter, the elements in combination cool and move toward their natural place in the cosmos, thereby bringing the organism to ruin.

The conceptual alignment of the two popular beliefs (that time is just celestial motion and also causes corruption) and the underlying philosophical truths (that time is related to celestial motion, one aspect of which is causally relevant to corruption) goes a long way toward explaining Aristotle's permissive attitude toward common speech habits regarding time. However, the explanation offered for seasonal growth and decay in terms of the sun's motion along the ecliptic is structurally symmetric; as Aristotle says, "the two processes occupy equal times" and are continuous. Why, then, is he content to approve of the common opinion that time is more properly the cause of deterioration rather than of improvement?

The answer to this question is to be found at a more fundamental level of Aristotle's account of time. Time, as we have seen, is a feature of motion, which in turn is analyzed partly in terms of telic properties – "time's arrow" is essentially a metaphysically gussied-up version of "motion's arrow." Since each of the four terrestrial elements has a type of motion that is natural to it, the progression of time will naturally track the motion of each elemental type toward its unique place in the cosmos. Thus, even without the change in temperature brought about by the sun's seasonal approach and recession, elements in combination are naturally disposed to dissociate. It is only by dint of some external agency (whether it be heat, psychic formation, or otherwise) that the elements are brought together to form mixtures and compounds.

So the reading of Aristotle's deflationary account of time's corrosive effects that was found to be wanting above is correct in at least this respect: the natural disposition of material objects – insofar as they are *material* objects – is to come apart. And since time is metaphysically derivative of motion, time is closely related to dissolution and decay.

All of this discussion of dissolution provides a convenient segue to our next topic: the puzzles of IV.10.

CHAPTER 13

Dissolving the puzzles of IV.10

The time has come for me to make good on the promise I issued in the Introduction. As I understand these puzzles, their purpose is to cast doubt on the very existence of time and to problematize the status of the now, and I believe that Aristotle clearly takes them to be sophistical.[1] There is in *Physics* IV no explicit, systematic treatment of them (nor anywhere else in the corpus, for that matter). Indeed, immediately after rehearsing the puzzles, Aristotle forthrightly drops them and takes up the other principal question concerning time, namely its nature (218a30). This suggests to me that the puzzles are easily solved, or so Aristotle thinks, when one sees the mistaken conception of time on which they rest and replaces that mistaken conception with the proper concept. And this is in fact the case. Once we get the puzzles in view, it will not be difficult to understand why Aristotle didn't trouble himself to dismantle them, since they aren't really the obstacles they appear to be, but rather are for the most part mirages.

I recast the puzzles here in the form of three arguments, the second offered as a response to an implicit objection to the first:

Argument (1): Any part of time one cares to consider consists exclusively of the past, the future, or both. But neither the past nor the future *is* (since the past *was*, and the future *will be*). So there is no part of time that *is*. Hence, time doesn't exist.

[Implicit objection: argument (1) fails to consider as a part of time the present instant – the now – which certainly *is*.]

Argument (2): If the now is a part of time, then either: (a) the now is capable of measuring time, or (b) time is composed of nows. But (c) the now is not capable of measuring time, and (d) no one thinks that time is composed of nows. So the now is not a part of time. Hence, argument (1) is vindicated.

Argument (3): The now must either be: (a) always different, or (b) always the same. If (a) the now is always different, then any particular now one cares to

[1] Hussey (1983), 140–1, speculates that Plato's *Prm.* 152b–c might have been the source of inspiration for these arguments. As we shall see below, one of the three arguments appears to have been lifted nearly verbatim from *Prm.* 156d ff.

consider must have ceased to be at some time, which must either be: (i) when it exists, or (ii) when a different now exists. But (iii) it's impossible that the now should cease to be when it exists, since doing so would prevent it from being at all. And (iv) it's impossible that the now should cease to exist when a different now exists, since doing so would require that it exist simultaneously with the infinitely many nows that lie between the two, which is absurd. So it's impossible that the now should always be different. But it's also impossible that (b) the now should always be the same. For (i) the now is a limit, and it's impossible that anything should exist with/as a single limit. And (ii) if the now is always the same, events taking place 10,000 years ago are simultaneous with those taking place today, which is absurd. Hence, the status of the now is highly problematic.

Let us examine the merit of each of these. Argument (1) purports to establish that time is unreal on the grounds that time consists exclusively of parts that are either past or future. As Miller notes, this argument (which he dubs "the whittling argument") is revived by Augustine in Book XI of his *Confessions*.[2]

But while Augustine may find the argument compelling, Aristotle surely does not. The puzzle monger treats an accidental characterization of time as an essential characterization, and then erroneously draws conclusions about the very existence of time based on this accidental characterization. Evidently, argument (1) presupposes that the proper sense of "time" is "that which consists exclusively of the past and the future." But this obviously cannot be the proper sense of the term, for even the puzzle monger will have to admit that the past and future are themselves

[2] Miller (1974), 133. Chapter 20 reads: "But even now it is manifest and clear that there are neither times future nor times past. Thus it is not properly said that there are three times, past, present, and future. Perhaps it might be said rightly that there are three times: a time present of things past; a time present of things present; and a time present of things future. For these three do coexist somehow in the soul, for otherwise I could not see them. The time present of things past is memory; the time present of things present is direct experience; the time present of things future is expectation. If we are allowed to speak of these things so, I see three times, and I grant that there are three. Let it still be said, then, as our misapplied custom has it: 'There are three times, past, present, and future.' I shall not be troubled by it, nor argue, nor object – always provided that what is said is understood, so that neither the future nor the past is said to exist now. There are but few things about which we speak properly – and many more about which we speak improperly – though we understand one another's meaning." (*Quod autem nunc liquet et claret, nec futura sunt nec praeterita, nec proprie dicitur: tempora sunt tria, praeteritum, praesens et futurum, sed fortasse proprie diceretur: tempora sunt tria, praesens de praeteritis, praesens de praesentibus, praesens de futuris. sunt enim haec in anima tria quaedam, et alibi ea non video praesens de praeteritis memoria, praesens de praesentibus contuitus, praesens de futuris expectatio. si haec permittimur dicere, tria tempora video fateorque, tria sunt. dicatur etiam: tempora sunt tria, praeteritum, praesens, et futurum, sicut abutitur consuetudo; dicatur. ecce non curo nec resisto nec reprehendo, dum tamen intellegatur quod dicitur, neque id quod futurum est esse iam, neque id quod praeteritum est. pauca sunt enim quae proprie loquimur, plura non proprie, sed agnoscitur quid velimus.*)

times, rendering the definition vacuous. So while it may be useful for certain purposes to suppose that time has as its "parts" the past and the future, it certainly is not the case that "the past and the future" qualifies as a satisfactory answer to the question of what time *is*, as it makes no headway on the issue whatsoever.[3] We have seen what Aristotle takes to be the proper sense of "time," and anyone who would deny that this exists would either have to deny that motion exists or that motion is perceptible. But whoever is willing to argue for either of these claims is, as Aristotle would say, "unable to distinguish what is self-evident from what is not … Presumably such a person must be talking about words without any thought to correspond" (193a5–9).[4]

We may note that the implicit objection to argument (1) is a red herring insofar as it accepts the specious characterization of time on which argument (1) is based. In spite of this fact, argument (2) is interesting insofar as it raises the question of the now's place in time. Aristotle agrees that the now is no part of time (220a18–19), but as we have seen, he is committed to the view that the now is capable of (indeed, necessary for) measuring time insofar as pairs of nows define intervals that function as the units of measurement. This, we should recall, is the basis for Aristotle's notion of "formal simultaneity," for which I provided an analysis in section II.3 above. But more to the immediate point, were there no now, there could be no time, since time is the measurable aspect of motion, and the now is required for the measurement of movements (220a1–4).

Argument (3) is the only puzzle truly deserving of a proper solution, and it echoes an argument in Plato's *Parmenides*, where the dialogue's namesake interrogates a young Socrates:

> "Is there, then, this strange thing in which it [the One] might be, just when it changes?"
>
> "What strange thing?"
>
> "The moment [*to exaiphnês*]. The moment seems to signify something such that changing occurs from it to each of two states. For a thing doesn't change from rest while rest continues, or from motion while motion continues. Rather, this strange creature, the moment, lies between motion and rest, being in no time at all, and both to it and from it the moving thing [*kinoumenon*] changes to resting and the resting thing changes to moving." (156d1–e3)[5]

[3] Compare: Aristotle would never regard the simple enumeration of parts as an adequate definition of any parted object, whether natural or artificial. Cf. *Tht*. 207c–208b.

[4] οὐ δυναμένου κρίνειν ἐστὶ τὸ δι᾽ αὑτὸ καὶ μὴ δι᾽ αὑτὸ γνώριμον … ὥστε ἀνάγκη τοῖς τοιούτοις περὶ τῶν ὀνομάτων εἶναι τὸν λόγον, νοεῖν δὲ μηδέν. Cf. *Phys*. 1.2, 184b25–185a17.

[5] Ἆρ᾽ οὖν ἔστι τὸ ἄτοπον τοῦτο, ἐν ᾧ τότ᾽ ἂν εἴη, ὅτε μεταβάλλει; – Τὸ ποῖον δή; – Τὸ ἐξαίφνης. τὸ γὰρ ἐξαίφνης τοιόνδε τι ἔοικε σημαίνειν, ὡς ἐξ ἐκαίνου μεταβάλλον εἰς ἑκάτερον. οὐ γὰρ

Aristotle is clearly sensitive to the peculiar status of the now, and we have already seen that he remarks in several places that there is a sense in which the now is the same, and there is a sense in which it is different: insofar as a now has a single kinetic cut (or more properly, a single kinetic world slice) as its material constituent, it is one; insofar as it is treated as a point of division within the movement, it is two.

But while this line of thought is familiar, it is unresponsive to the challenge. The argument demands that Aristotle embrace the claim that the now is always different (*aiei heteron*) or the claim that it is always the same (*aiei to auto*). The dilemma alleges that embracing the second alternative commits one to an "eternalist" conception of time, according to which "events taking place 10,000 years ago are simultaneous with those taking place today." Since I have argued that Aristotle is an *A* theorist, there had better be decisive evidence that he rejects the second alternative, either by embracing the first alternative or by finding a third way.

And indeed there is such evidence. Aristotle grasps the first horn of the dilemma: he affirms that the now is always different, but he denies that this entails that the now "must have ceased to be at some time." We examined passages from *Physics* VIII.8 in section 6.2, where the challenge was understanding how conceptual division of movements occurs. Within the context of that issue, Aristotle provides the basis for his response to the dilemma posed in argument (3):

> When its motion is continuous, *A* cannot either have come to be or have ceased to be at the point [of division] *B*; it can only have been there at a now, and not in any time except the whole of which the now is a division … it is there only in a cut of time [*en tomê chronou*], not for any time [*ouk en chronô*]. (262a28–b22)[6]

So Aristotle simply denies that there is any answer to the questions "when does the *kinoumenon* come to be at *B*?" and "when does it cease to be there?" other than "there is no time when it does so." This result follows immediately from two of his other commitments: (i) his conception of time as measured bit of motion, and (ii) his conception of being in time as having a primary time.

ἔκ γε τοῦ ἑστάναι ἑστῶτος ἔτι μεταβάλλει, οὐδ' ἐκ τῆς κινήσεως κινουμένης ἔτι μεταβάλλει· ἀλλὰ ἡ ἐξαίφνης αὕτη φύσις ἄτοπός τις ἐγκάθηται μεταξὺ τῆς κινήσεώς τε καὶ στάσεως, ἐν χρόνῳ οὐδενὶ οὖσα, καὶ εἰς ταύτην δὴ καὶ ἐκ ταύτης τό τε κινούμενον μεταβάλλει ἐπὶ τὸ ἑστάναι καὶ τὸ ἑστὸς ἐπὶ τὸ κινεῖσθαι.

6 ὅταν δὲ συνεχῶς φέρηται, οὔτε γεγονέναι οὔτε ἀπογεγονέναι οἷον τε τὸ Α κατὰ τὸ Β σημεῖον, ἀλλὰ μόνον εἶναι ἐν τῷ νῦν, ἐν χρόνῳ δ' οὐδενὶ πλὴν οὗ τὸ νῦν ἐστιν διαίρεσις, ἐν τῷ ὅλῳ … ἀλλ' ἦν ἐν τομῇ χρόνου καὶ οὐκ ἐν χρόνῳ.

With respect to (i), we have seen (in section 5.4) that kinetic cuts are not themselves movements, and a single point has no measure.

With respect to (ii), we have also seen (in section 11.2) that Aristotle regards having a primary time as a necessary condition for being in time. Primary time is defined in terms of pairs of kinetic world slices, and condition (a) in the definiens requires that $S_2 > S_1$, thereby preventing any single kinetic world slice (or kinetic cut) from having its own primary time. Aristotle's remark that *A* cannot have been at *B* "in any time except the whole of which the now is a division" is intended precisely to deny that there is any primary time for *A*'s being at *B*.

But since a kinetic cut is the material constituent of a now, the fact that a kinetic cut has no primary time entails that a now has none, either. So to the question "when does the now cease to be?" Aristotle must answer "there is no such time." And for this reason, he can embrace the claim that the now is always different without embarrassment, and argument (3) goes the way of the other two sophistical arguments.

To dispatch such seemingly deep puzzles as quickly as we have is perhaps less gratifying than straining to discover their solutions. But when one's target is a paper tiger, it is not only unnecessary, but also foolish, to enlist the aid of a safari guide and to secure actual rifles. I take it that this is the explanation for Aristotle's failure to say very much about the puzzles in the course of developing his theory of time: once we see what his view is, the puzzles (to a large extent) simply fall apart.

CHAPTER 14

Concluding summary and historical significance

Both Plato and Aristotle were working against the background of an entrenched popular conception of time when developing their philosophical theories. Hesiod couches his advice regarding the timing of works in terms of celestial events, and the type of chronometer used by the ancients to keep "natural time" (namely, the sundial) functions by tracking the movement of the most prominent heavenly body. The lack of a commonly recognized epoch meant that the calendric conventions employed by common folks and historians alike named years by the number of years removed from some salient event. I argued in Chapter 2 that in the *Timaeus* Plato accomodates this traditional conception of time to a fairly high degree by defining time in terms of the regular motion of the celestial bodies – that is, as "motion-according-to-number." Such motion, on Plato's view, is the product of the demiurge's imposition of numerical form onto what were previously disorderly movements:

> In this way and for these reasons night-and-day, the period of a single circling, the wisest one, came to be. A month has passed when the moon has completed its own cycle and overtaken the Sun; a year when the Sun has completed its own cycle.
>
> As for the periods of the other bodies, all but a scattered few have failed to take note of them. Nobody has given them names or investigated their numerical measurements relative to each other. And so people are all but ignorant of the fact that time really is the wanderings of these bodies, bewilderingly numerous as they are and astonishingly variegated. (39c)[1]

The demiurge created time by crafting a cosmic-scale clockwork out of the raw material (in this case, motion) that he found in the receptacle.

[1] νὺξ μὲν οὖν ἡμέρα τε γέγονεν οὕτως καὶ διὰ ταῦτα, ἡ τῆς μιᾶς καὶ φρονιμωτάτης κυκλήσεως περίοδος· μεὶς δὲ ἐπειδὰν σελήνη περιελθοῦσα τὸν ἑαυτῆς κύκλον ἥλιον ἐπικαταλάβῃ, ἐνιαυτὸς δὲ ὁπόταν ἥλιος τὸν ἑαυτοῦ περιέλθῃ κύκλον. τῶν δ' ἄλλων τὰς περιόδους οὐκ ἐννενοηκότες ἄνθρωποι, πλὴν ὀλίγοι τῶν πολλῶν, οὔτε ὀνομάζουσιν οὔτε πρὸς ἄλληλα συμμετροῦνται σκοποῦντες ἀριθμοῖς, ὥστε ὡς ἔπος εἰπεῖν οὐκ ἴσασιν χρόνον ὄντα τὰς τούτων πλάνας, πλήθει μὲν ἀμηχάνῳ χρωμένας, πεποικιλμένας δὲ θαυμαστῶς·

By regularizing and fitting together the diverse movements of the celestial bodies, the demiurge created a synchronized system that not only grounds our temporal judgments, but whose motion just is time.

Aristotle agrees with Plato that the celestial clock is the ultimate standard for all temporal judgments, but he denies that time consists in nothing more than the regular motion of heavenly bodies. For first of all, Aristotle cannot subscribe to Plato's view as offered, because he rejects Plato's claim that orderly celestial motion is the result of a creative act of divine intelligence. Aristotle regards the heavenly bodies as ungenerated objects composed of pure and incorruptible *aether*. Given that the natural motion of *aether* is circular, he believes that the stars and planets have always moved in their orbits with perfect regularity.

More importantly, he offers in IV.10 what he takes to be decisive arguments against strictly identifying time with motion. So time is not any particular variety of motion, much less motion brought into being by a divine craftsman. But time is not unrelated to motion, either. This is evident when we consider the fact that we notice time exactly when we notice motion. The best explanation for this fact, Aristotle argues, is that time is an evident proper feature of motion – an aspect of motion (and of nothing else, except by way of its relation to motion) that one cannot fail to perceive if she perceives movement *as* a movement. Thus, Plato's identification of time with motion was in error, but it was a forgivable error and one that Aristotle regarded as helpful in discovering the genuine nature of time.

Plato believed that motion actually existed prior to time. Aristotle rejects this position but does endorse the weaker view that motion *could* exist without time if there were no counting souls. He reasons that, as a perceptible feature of motion, time would not exist if there were no perceivers, even if motion remained in existence. *Being perceptible* is a modal relational property conferred upon motion by the existence of perceivers – as he believes, relational properties do not exist independently of their relata.

The pair <time, motion> is included in the extension of the evident perceptible feature relation because time is the measurability (or a measure) of motion. This view is consistent with the common beliefs and calendric conventions of Aristotle's day, and provides an opportunity to elaborate upon Plato's view of time as the product of imposing form onto motion as matter.

Considered in themselves, movements are continuous, partless wholes. Motion is the actuality of a *kinoumenon*, an accidental compound whose matter is (typically) an ordinary substance and whose form is (always) a telic property. State-properties of *kinoumena* form the basis for kinetic cuts, which are the kinetic analogue of geometric points and instants in time. By virtue of this ontological structure and the capacities of percipient agents, motion is susceptible to partitioning into undetached parts of arbitrary length. And this susceptibility is just what time is. Motion stands as the matter of time, indeterminate with respect to quantity. Perception provides the form of time by conferring upon motion the relational property of determinability and sometimes actually making movements determinate with respect to quantity.

This feature of Aristotle's view is perhaps most evident when considering the ultimate standards of temporal measurement – the several movements of the heavenly bodies and, most prominently, of the sun – which are both continuous and eternal. The standard units of time (days, months, years) simply would not exist if there were no rational souls in the world. Not merely because there would be no common names for the periods, but because in order for such a period to exist, the continuous movement of which it is a part must be divided by identifying a kinetic cut (specified by some azimuth reading, for example) from which the star departs and once again returns. The kinetic cut exists independently of our individuation of it, but in order for it to function as a point of demarcation, it must be a point of division; and in order for it to be a point of division, it must be treated in a certain way by rational agents, namely in its dual aspect as the end of the earlier segment and beginning of the later segment.

Both Plato and Aristotle take space to be metaphysically more basic than time. For Plato, the receptacle is one of the three primitive features of the universe, coordinate with being and becoming in fundamentality. The once-erratic movements taking place in space were marshaled by the demiurge into a mathematically structured system of rational movements, thus bringing into existence time. Aristotle likewise takes spatial magnitude to be a basic feature of the world. Some telic properties are spatial in character, and we percipients mark *kinoumena* out as the kinetically extended things they are, comparing them against one another (and against the universal standard: the movement of the sun) in the process. Such an asymmetry between space and time is rejected by modern physics, but for philosophers trying to accommodate a deeply entrenched view of time as being somehow connected with celestial motion, the asymmetry is not surprising.

As we saw in Chapter 2, Plato is committed to the view that temporal extension existed prior to the genesis of time. It is on these grounds that we were justified in saying that temporal extension is not part of Plato's conception of time, properly speaking, though it is metaphysically necessary for time's existence. In conceiving time as the periodic motion of the celestial spheres, Plato conceives time to be, first and foremost, periodicity. But because the distinction between past and present is parasitic upon the existence of a well-defined present – whose existence requires a well-defined simultaneity relation – genuine temporal passage was also created by the demiurge when he worked the chaotic movements into a synchronized system.

By defining time as the measurability (or as a measure) of motion, Aristotle's concept of time also places greatest emphasis on periodicity. Also like Plato, Aristotle regards a certain kind of extension as being necessary for the existence of time. But whereas Plato has very little to say about the nature of pre-creation motion, Aristotle has a highly sophisticated account of the nature and structure of motion. Indeed, Aristotle's theoretically fertile account of motion is precisely what provides the means for him to account for periodicity: measuring motion involves "carving up" kinetic extension by means of treating a single kinetic cut as a point of division.

Further, while the argument I have attributed to Plato to the conclusion that the demiurge created a well-defined simultaneity relation is only suggestive at best, Aristotle has available (on the hylomorphic interpretation) the theoretical resources to define several distinct, interesting, and useful simultaneity relations. Indeed, formal simultaneity (which Aristotle calls "unity in form") provides the essential content of Aristotle's idea that time is a measure of motion.

Aristotle's motive for developing a hylomorphic account of time seems to have been twofold. First, he no doubt wished to retain to the greatest extent possible the well-established conception of time as being somehow intimately related to the movement of celestial bodies and as involving numbering in some way. Second, he must have felt compelled to account for time's existence in the way that he thought the existence of all non-substances is to be accounted for: by tracing its existence to that of substances and their features. But to complicate matters, Aristotle recognized that the business of keeping time is largely conventional, and that our experience of time is deeply subjective. For these reasons he devised an account of time according to which its existence is partly dependent upon the activities of percipient agents. The result, as I hope I have shown, is a sophisticated and interesting view that manages to avoid the most damning objection that plagues contemporary views of time, while still preserving – and, indeed, accounting for– our intuition that time in some sense passes by locating the basis for temporal passage within "kinetic passage," which is something that can be analyzed in temporal-free terms.

Does Aristotle have anything to teach contemporary theorists working in the philosophy of time? Perhaps. It is certainly the case, however, that current thinkers enjoy the benefits (or suffer the burdens) of a vastly richer body of knowledge and opinions on which to base their accounts. If anything, Aristotle's theory of time provides a promising model for a coherent account of temporal passage precisely because it is immune to the sorts of objections raised by McTaggart, Dummett, and Williams. But grounded as it is on his idiosyncratic account of motion and change, the theory is precarious, to say the least. For to the extent that Aristotle's theory of motion is unacceptable to anyone who rejects telic properties (and surely that camp is in the majority), any attempt to adapt Aristotle's temporal theory will have to provide a suitable surrogate account of change.

One natural place to look for such a surrogate is toward causation. If one can provide an account of change in terms of causation, and if that account is free of temporal terms, then the possibility of founding time on motion remains open. In fact, this strategy seems to me to be somewhat promising, since simultaneity is already defined within the special theory of relativity in terms of possible causal relations between objects. Any such causal account of change would have to incorporate some version of the telic aspect of Aristotle's view in order to be suited to serve its theoretical role, and indeed some philosophers have maintained that the necessary asymmetry of causal relations is precisely what fixes the necessary, intrinsic direction of the temporal order.[2]

A temporal-free account of causation might be framed in terms of counterfactual analysis, since the latter is quite popularly taken to be a special case of possible-worlds analysis, and neither modal realists nor conventionalists posit any temporal relations between diverse possible worlds.[3] Simultaneity and temporal priority and posteriority obtain only between events occurring within the same world.

I am impressed by the fact that David Lewis pursued this line of thought on several occasions.[4] And although he makes a case for counterfactual

[2] See, e.g., Grünbaum (1963).

[3] See, e.g., Forbes (1993, 92) who unwittingly shows his Aristotelian roots when he says that "there is nothing intrinsically perverse about beginning with comparative-magnitude facts and using these to generate instants and the intervals between them, rather than the other way round." See also Lewis (1986), 71.

[4] Lewis (1973), (1979), (1986), 23.

temporal asymmetry, he admits that he doesn't know how to connect this with the asymmetry of temporal passage – or as he puts it, with the "asymmetry of entropy."[5] Perhaps Lewis never harbored hopes of making such a connection, and his admission might be enough to convince some of his successors that the project described is unworkable. Nevertheless, the fact that central elements of Aristotle's view are cropping up nowadays in a new guise inclines me to think that cleverer philosophers than I might well see their way to fashioning a viable, genuinely Aristotelian account of time.

Whether such possibilities are ever realized or not, I am hopeful that Aristotle's discussion of time will be spared the unwarranted criticism that it has received for far too long from far too many good philosophers. If this book contributes in any way to that result, I should have to consider it a success.

[5] Lewis (1979), 476.

References

Annas, Julia. 1975. "Aristotle, Number and Time." *Philosophical Quarterly* 25 (April): 97–113.

1992. "Aristotle on Memory and the Self." In Nussbaum and Rorty, eds., 297–312.

Anscombe, G. E. M. 1953. "The Principle of Individuation." *Proceedings of the Aristotelian Society, Supplementary Volumes* 27: 83–96.

1956. "Aristotle and the Sea Battle." *Mind* 65 (257): 1–15.

Audi, Robert, ed. 1995. *The Cambridge Dictionary of Philosophy*. Cambridge University Press.

Augustine, St. 1980. *Confessions*, ed. J. Gibb and W. Montgomery. New York: Garland.

Barnes, Jonathan, ed. 1984. *The Complete Works of Aristotle*. Princeton University Press.

Bilmes, J. A. 1993. "Timing is of the Essence: Perceptual and Computational Techniques for Representing, Learning, and Reproducing Expressive Timing in Percussive Rhythm." Master's thesis, Massachusetts Institute of Technology.

Bogen, James. 1991. "Aristotelian Contraries." *Topoi* 10: 53–66.

Bolotin, David. 1997. "Aristotle's Discussion of Time: An Overview." *Ancient Philosophy* 17: 47–62.

Bostock, David. 1980. "Aristotle's Account of Time." *Phronesis* 25: 148–69.

1991. "Aristotle on Continuity in *Physics* VI." In L. Judson, ed., *Aristotle's Physics: A Collection of Essays*, 179–212. Oxford: Clarendon Press.

2006. *Space, Time, Matter, and Form: Essays on Aristotle's Physics*. Oxford: Clarendon Press.

Broad, C. D. 1967. "Ostensible Temporality." In Richard M. Gayle, ed., *The Philosophy of Time*, 117–42. Garden City, NY: Anchor.

Broadie, Sarah. 1982. *Passage and Possibility*. Oxford: Clarendon Press.

1983. "Instants of Motion in Aristotle's *Physics* VI." *Archive für Geschichte der Philosophie* 65: 128–46.

1984. "Aristotle's Now." *Philosophical Quarterly* 34 (April): 104–28.

Brumbaugh, Robert S. 1966. *Ancient Greek Gadgets and Machines*. New York: Crowell.

1980. "Time Passes: Platonic Variations." *Review of Metaphysics* 33 (June): 711–26.

Burnyeat, Myles. 1992. "Is an Aristotelian Philosophy of Mind Still Credible? A Draft." In Nussbaum and Rorty, eds., 15–26.

Caston, Victor. 1998. "Aristotle and the Problem of Intentionality." *Philosophy and Phenomenological Research* 53 (June): 249–98.

2002. "Aristotle on Consciousness." *Mind* 111 (444): 751–815.

Cicero, Marcus Tullius. 1970. *De Oratore*, trans. J.S. Watson. Carbondale: Southern Illinois University Press.

Cohen, S. M. 2008. "Kooky Objects Revisited: Aristotle's Ontology." *Metaphilosophy* 39 (January): 3–19.

Conen, Paul F. 1952. "Aristotle's Definition of Time." *New Scholasticism* 26 (October): 441–58.

Coope, Ursula. 2001. "Why Does Aristotle Say That There Is No Time without Change?" *Proceedings of the Aristotelian Society* 101: 359–67.

2005. *Time for Aristotle: Physics* IV.*10–14*. Oxford: Clarendon Press.

Cooper, John. 1986. *Reason and Human Good in Aristotle*. Indianapolis: Hackett.

Cooper, John, ed. 1997. *Plato: Complete Works*. Indianapolis: Hackett.

Corish, Denis. 1969. "The Continuum." *Review of Metaphysics* 22 (3): 523–46.

1976. "Aristotle's Attempted Derivation of Temporal Order from That of Movement and Space." *Phronesis* 21: 241–51.

Cornford, F. M. 1937. *Plato's Cosmology*. London: Routledge and Kegan Paul.

De Tollenaere, M. 1961. "Aristotle's Definition of Time." *International Philosophical Quarterly* 1 (September): 453–67.

Diels, Hermann. 1912. *Die Fragmente der Vorsokratiker*, 1st edn. Berlin: Weidmann.

Dummett, Michael. 1978. "A Defence of McTaggart's Proof of the Unreality of Time." In *Truth and Other Enigmas*, 351–7. Cambridge, MA: Harvard University Press.

Earman, John and R. M. Gale. 1995. "Time". In Audi, ed.

Edwards, Paul. 1949. "Necessary Propositions and the Future." *Journal of Philosophy* 46: 155–7.

Evans, James. 1998. *The History and Practice of Ancient Astronomy*. New York: Oxford University Press.

Everson, Stephen. 1997. *Aristotle on Perception*. Oxford: Clarendon Press.

Fitzgerald, Paul. 1969. "The Truth about Tomorrow's Sea Fight." *Journal of Philosophy* 66 (June): 307–29.

Forbes, Graham. 1993."Time, Events, and Modality." In Robin Le Poidevin and Murray MacBeth, eds., *The Philosophy of Time*, 80–95. New York: Oxford University Press.

Gould, Josiah B. 1993. "Aristotle on Time and Possibility in *De Caelo* 1.12." *Philosophical Inquiry* 15: 59–74.

Graham, Daniel W. 1988. "Aristotle's Definition of Motion." *Ancient Philosophy* 8: 209–15.

Grünbaum, Adolf. 1963. *Philosophical Problems of Space and Time.* New York: Alfred A. Knopf.

Guthrie, W. K. C. 1962. *A History of Greek Philosophy*, vol. 1. London: Cambridge University Press.

Herodotus. 1947. *The History of Herodotus*, trans. George Rawlinson. New York: Tudor Publishing.

Hesiod. 1964. *Works and Days*, trans. Hugh G. Evelyn-White. Cambridge, MA: Harvard University Press.

Hintikka, Jaakko. 1957. "Necessity, Universality, and Time in Aristotle." *Ajatus* 20: 65–90.

1973. *Time and Necessity.* Oxford: Clarendon Press.

Hintikka, Jaakko, Knuuttila Simo, and Remes Unto. 1977. "Aristotle on Modality and Determinism." *Acta Philosophica Fennica* 29: 13–58.

Hussey, Edward. 1983. *Aristotle's Physics, Books III and IV.* Oxford: Clarendon Press.

Inwood, Michael. 1991. "Aristotle on the Reality of Time." In L. Judson, ed., *Aristotle's Physics: A Collection of Essays*, 151–78. Oxford: Clarendon Press.

Janich, Peter. 1985. *Protophysics of Time.* Boston Studies in the Philosophy of Science 30. Dordrecht: Reidel.

Johnston, A. 2007. "The Simple Bodies As Unities of Quantity and Quality in Aristotle's *On Generation and Corruption.*" In Michael Treschow, Willemien Otten, and Walter Hannam, eds., *Divine Creation in Ancient, Medieval, and Early Modern Thought: Essays Presented to the Revd. Dr. Robert D. Crouse.* Brill's Studies in Intellectual History 151, 73–84. Leiden: Brill.

Kahn, Charles H. 1992. "Aristotle on Thinking". In Nussbaum and Rorty, eds., 359–79.

Kant, Immanuel. 1965. *Critique of Pure Reason*, trans. Norman Kemp Smith. New York: St. Martin's Press.

Keil, B. 1902. *Anonymus Argentinensis.* Strasburg: Trübner.

King, Hugh R. 1949. "Aristotle and the Paradoxes of Zeno." *Journal of Philosophy* 46 (October): 657–70.

Kirk, G. S. and J. E. Raven. 1957. *The Presocratic Philosophers.* Cambridge University Press.

Kosman, L. A. 1969. "Aristotle's Definition of Motion." *Phronesis* 14: 40–62.

1975. "Perceiving That We Perceive." *Philosophical Review* 84: 499–519.

1992. "What Does the Maker Mind Make?" In Nussbaum and Rorty, eds., 343–58.

Kretzmann, Norman. 1976. "Aristotle on the Instant of Change (II)." *Proceedings of the Aristotelian Society, Supplementary Volumes* 50: 91–114.

Lear, Jonathan. 1981. "A Note on Zeno's Arrow." *Phronesis* 26: 91–104.

Lewis, David. 1973. "Causation." *Journal of Philosophy* 70: 556–67.

1979. "Counterfactual Dependence and Time's Arrow." *Noûs* 13: 455–76.

1986. *On the Plurality of Worlds.* Oxford: Blackwell.

McClelland, Richard T. 1981. "Time and Modality in Aristotle, *Metaphysics* IX.3–4." *Archiv für Geschichte der Philosophie* 63: 130–49.

McKirahan, Richard D. 1994. *Philosophy Before Socrates*. Indianapolis: Hackett.
McTaggart, J. M. E. 1927. *The Nature of Existence*, vol. II. Cambridge University Press.
Mellor, D. H. 1981. *Real Time*. Cambridge University Press.
Miller, Fred. 1974. "Aristotle on the Reality of Time." *Archive für Geschichte der Philosophie* 56: 132–55.
Modrak, Deborah K. W. 1987. *Aristotle: The Power of Perception*. University of Chicago Press.
Mohr, Richard D. 1986. "Plato on Time and Eternity." *Ancient Philosophy* 6: 39–46.
Nagel, Thomas. 1986. *The View from Nowhere*. New York: Oxford University Press.
Newton-Smith, W. H. 1986. "Space, Time and Space-Time: A Philosopher's View". In Raymond Flood and Michael Lockwood, eds., *The Nature of Time*, 22–35. Oxford: Blackwell.
Nussbaum, Martha C. and Amélie Oksenberg Rorty, eds. 1992. *Essays on Aristotle's De Anima*. Oxford: Clarendon Press.
Owen, G. E. L. 1966. "Plato and Parmenides on the Timeless Present." *Monist* 50 (July): 317–40.
1975a "Aristotle, Number, and Time." *Philosophical Quarterly* 25 (April): 97–113.
1975b."'*Tithenai ta Phainomena*'." In Jonathan Barnes, Malcolm Schofield, and Richard Sorabji, eds., *Articles on Aristotle, vol. I, Science*, 113–26. London: Duckworth.
1975c. "Zeno and the Mathematicians." In R. E. Allen and D. J. Furley, eds., *Studies in Presocratic Philosophy*, vol. II, 143–65. London: Routledge and Kegan Paul.
1976. "Aristotle on Time." In Peter Machamer and Robert Turnbull, eds., *Motion and Time, Space and Matter: Interrelations in the History of Philosophy and Science*, 3–27. Columbus: Ohio State University Press.
1986. "Aristotle: Method, Physics and Cosmology." In Martha Nussbaum, ed., *Logic, Science, and Dialectic*, 151–64. Ithaca, NY: Cornell University Press.
Rhodes, P. J. 1981. *A Commentary on the Aristotelian Athenaion Politeia*. Oxford: Clarendon Press.
Roark, Tony. 2003. "Aristotle's Definition of Time Is Not Circular." *Ancient Philosophy* 23: 301–18.
2004. "Why Aristotle Says That There Is No Time without Change." *Apeiron* 37: 227–46.
2005. "Aristotelian Temporal Passage." *Philosophical Writings* 28: 23–33.
2009. Review of *Time for Aristotle*, Ursula Coope. *Mind* 118 (470): 459–62.
Ross, G. R. T. 1973. *Aristotle: De Sensu and De Memoria*. New York: Arno Press.
Ross, W. D. 1936. *Aristotle's Physics*. Oxford: Clarendon Press.
Russell, Bertrand. 1903. *The Principles of Mathematics*. New York: Norton.
Salmon, Wesley. 1980. *Space, Time and Motion: A Philosophical Introduction*. Minneapolis: University of Minnesota Press.

Savitt, Steven F. 2001. "A Limited Defense of Passage." *American Philosophical Quarterly* 38: 261–70.

Shoemaker, Sydney. 1969. "Time without Change." *Journal of Philosophy* 66 (June): 363–81.

Smith, John E. 1969. "Time, Times, and the 'Right Time'; *Chronos* and *Kairos*." *Monist* 53: 1–13.

Sorabji, Richard. 1972. "Aristotle, Mathematics, and Colour." *Classical Quarterly* 22: 293–308.

1976. "Aristotle on the Instant of Change (I)." *Proceedings of the Aristotelian Society, Supplementary Volumes* 50: 69–89.

1983. *Time, Creation, and the Continuum*. London: Duckworth.

1992. "Intentionality and Physiological Processes: Aristotle's Theory of Sense Perception." In Nussbaum and Rorty, eds., 195–226.

Taylor, A. E. 1928. *A Commentary on Plato's Timaeus*. Oxford Clarendon Press.

Urmson, J. O., trans. 1992. *On Aristotle's Physics 4.1–5, 10–14*. Ithaca, NY: Cornell University Press.

Vlastos, Gregory. 1965. "The Disorderly Motion in the 'Timaeus'." In R. E. Allen, ed., *Studies in Plato's Metaphysics*, 379–99. London: Routledge and Kegan Paul.

1975a. "A Note on Zeno's Arrow." In R. E. Allen and D. J. Furley, eds., *Studies in Presocratic Philosophy*, vol. II, 184–200. London: Routledge and Kegan Paul.

1975b. "Zeno's Race Course." In R. E. Allen and D. J. Furley, eds., *Studies in Presocratic Philosophy*, vol. II, 201–20. London: Routledge and Kegan Paul.

Von Leyden, W. 1964. "Time, Number, and Eternity in Plato and Aristotle." *Philosophical Quarterly* 14 (January): 35–52.

White, Michael J. 1980. "Necessity and Unactualized Possibilities in Aristotle." *Philosophical Studies* 38: 287–98.

1981. "Fatalism and Causal Determinism: An Aristotelian Essay." *Philosophical Quarterly* 31: 231–41.

1989. "Aristotle on 'Time' and 'A Time'." *Apeiron* 22 (3): 207–24.

Wieland, W. 1962. *Die aristotelische Physik*. Göttingen: Vandenhoeck & Ruprecht.

Williams, D. C. 1951. "The Myth of Passage." *Journal of Philosophy* 47 (15): 457–72.

Zeyl, Donald J., trans. 1997. *Timaeus*. In Cooper, ed., 1224–92.

Index locorum

General index